M000202156

ENERGIZE!

ALSO BY MICHAEL BREUS, PHD

The Power of When: Learn the Best Time to do Everything

The Sleep Doctor's Diet Plan: Lose Weight Through Better Sleep

Beauty Sleep: Look Younger, Lose Weight, and Feel Great Through Better Sleep

ALSO BY STACEY GRIFFITH

*Two Turns from Zero: Pushing to Higher Fitness Goals —
Converting Them to Life Strength*

ENERGIZE!

Go from Shattered to Smashing It in 30 Days

MICHAEL BREUS, PhD

and

STACEY GRIFFITH

Vermilion
LONDON

1

Vermilion, an imprint of Ebury Publishing,
20 Vauxhall Bridge Road,
London SW1V 2SA

Vermilion is part of the Penguin Random House group of companies
whose addresses can be found at global.penguinrandomhouse.com

Copyright © Dr Michael Breus and Stacey Griffith 2021

Michael Breus, PhD and Stacey Griffith have asserted their right to be identified as the authors
of this Work in accordance with the Copyright, Designs and Patents Act 1988

First published in 2021 in Great Britain by Vermilion
First published in 2021 in the United States by Little, Brown Spark,
part of Hachette Book Group, New York

www.penguin.co.uk

A CIP catalogue record for this book is available from the British Library

ISBN 9781785043659

Printed and bound in Great Britain by Clays Ltd, Elcograf S.p.A.

The authorised representative in the EEA is Penguin Random House Ireland,
Morrison Chambers, 32 Nassau Street, Dublin D02 YH68

Penguin Random House is committed to a
sustainable future for our business, our readers
and our planet. This book is made from Forest
Stewardship Council® certified paper.

I want to dedicate this book in three ways:

*First, to my incredible family, Lauren, Cooper, and Carson,
as well as our animals, who give so much joy — Monty and Hugo,
aka Sugar Bear. Book 4, still going strong.*

*Second, a special dedication to Drs. Michael and Bridget Stamos — you
were right, all I needed to do was "**Wait for it**," be patient, and have
more balance in my life. Thank you seems trivial but truthful.*

*Finally, to all my patients throughout my more than twenty years
of practice. I love learning from you all, every time we meet.*

— MB

*I want to dedicate this book to my superstar of a grandmother,
Stella Mae Bjornson, whom I lost in March 2020, the same week the
Covid-19 pandemic shut everything down. Saying goodbye to a loved
one over FaceTime is a story that many of us have to tell in the world
we live in today. Grammy, I love you, and you would have
loved this book; it's prescriptive and to-the-point, as you always
told me to be in my fitness career.*

Mom, you're always my #1!

*MDSG, I LOVE YOU — thanks for always being
real with me, forever . . .*

— SG

Contents

Contents

ENERGIZE!

Our Lightning Strike

The two inspirations for this book were a cardiac event and a personal crisis.

MICHAEL'S STORY

On a balmy California evening a few years ago — I was forty-nine at the time — my wife and I were out to dinner with two friends who, luckily as it turned out, were both physicians. After a trip to the bathroom, I returned to the table, feeling strange. When I sat back down, I tried to tune in to one of my companions who was telling a story, but all I could hear were muffled nonsense words, like the man was speaking in a foreign language in slow motion.

Something's up, I thought.

All of a sudden, my peripheral vision started darkening around the edges. The voices got even weirder, and I thought I was having either a stroke or a heart attack. I broke out in a cold sweat, turned to my wife to say, "I don't feel so . . ."

And that was the last thing I remembered. I fell out of my chair and, thanks to the quick reflexes of my friend who caught me on the way down, I didn't smash my skull on the tile floor of the restaurant.

After about thirty seconds, I regained consciousness. My wife was looking down at me. She later recalled, "His face was as white as his teeth. I was terrified he was dead." One of my doctor friends was cradling my head, saying, "Michael, come back. Michael, come back."

From the floor, I realized that I'd been unconscious and had come to, but there was a big problem. My eyes were open, but everything was black. I whispered, "I can't see," and I think I started to cry.

My friend said, "Wait for it, just wait for it," and then, in a stunning flash, my vision snapped back, which was a huge relief, followed immediately by a powerful urge to throw up.

While I was "out," my wife had called 911. Two paramedics were already on the scene (we were dining at a food festival, and fortunately an ambulance was parked just down the street). One of the paramedics ripped my shirt open and put electrodes on my chest to measure my heart activity.

I said, "I think I'm going to vomit. I've got to turn over. Can somebody turn me over?" The paramedics started to flip me on my side . . . and then I had *another* blackout, the second heart stoppage in just several minutes.

When I regained consciousness the second time, I was on a gurney in the ambulance, speeding to the hospital. My wife was in the front seat with the driver, telling him how to get there faster (true story). As the paramedics wheeled me into the ER, a nurse appeared to hook me up to an IV. I'm not a huge fan of needles and objected, but the nurse grabbed my arm and just jammed it in . . . which triggered a *third* cardiac event in less than an hour. I passed out again. It was a long night.

The good news was that my heart activity was being monitored on an EKG by then, so the ER doctors could see what was going on. They discovered that I was dropping "P waves." In the heartbeat, the P wave is the ignition. You know when you put the key in the ignition of your car and it doesn't turn over, but you pump the gas anyway and flood the engine? My heart was that flooded engine.

After a night in the hospital, my doctors and I theorized that I might have a genetic condition. My father had had significant cardiac issues, after all. The first thing I did when I got out of the hospital was call my friend and colleague Mehmet Oz, aka Dr. Oz, a cardiothoracic surgeon. Dr. Oz referred me to the top electrophysiologist in the country at the UCLA Medical Center.

I arrived at my appointment with the heart specialist, fully expecting to be told that I'd need a pacemaker like my father. But after some diagnostic testing, the doctor said, "Michael, when you were in my waiting room, what was the average age of the people in there?"

"I'd say probably sixty-five." Sixteen years older than I was then.

"You're right. You're the healthiest person to come into this office in ten years," he said. "I don't want to put a pacemaker in you. It's the most overprescribed surgery there is. You need to make lifestyle changes. I think this had a lot to do with anxiety and not taking care of yourself. You really need to look at your stress levels."

What the . . . ? But I was in great shape! At a healthy weight, I ran three 5Ks a week and took supplements for my health. I was living what anyone would consider an objectively healthy lifestyle. It wasn't like I was doing anything *wrong*. I wasn't a morbidly obese chain-smoking alcoholic. I was probably more physically fit than I'd ever been.

And yet, in this "great shape," I'd had three cardiac events in a row that gave my wife and me (and our two kids) one of the most frightening nights of our lives. For the most part, I'd been doing what anyone should do for good health. But clearly, what worked for "anyone" wasn't cutting it for me. I would have to reevaluate my approach to my own fitness and health and make some changes or I might have another cardiac event. According to my wife, this was not an option.

Yup, the Sleep Doctor got a wake-up call.

So I pushed the pause button on my routine to figure out the best approach *for me,* since what I was currently doing had put me in the hospital.

STACEY'S STORY

Meanwhile, in New York City, I was having a crisis of a different kind. Often after teaching five classes in a day, I somehow had to rally to enjoy a night out with my partner and our close friends. I felt bad that I always seemed to be the odd man out, having to leave early to go to bed. Some of my mornings began at 6:00 a.m. at SoulCycle, where, as one of the founding master instructors, I was compelled to teach high-quality classes more than twenty times per week. Getting less than eight hours of sleep per night was like torture for me. As much as I tried to muster the strength to hang with my crew, it was hard for me to squeeze out what little energy I had left by the end of my day.

Everybody needs sleep. Granted, some might need more than others, but I knew I required a lot. Every *thing* needs to rest. Even your computer and phone must chill in their docks to recharge. Humans are no different.

It got to the point where I went to see my doctor about getting a prescription to help me maximize my day. At the end of the visit, she sent me home with a prescription to sleep more. Unfortunately, that was still impossible. For a while, I tried to just live on a different schedule than everyone else. But that wasn't working either. When I felt run-down, I would tend to slip into an argumentative state of mind. I'd be at a function or gathering and, as the hours passed, I would start looking at my watch anxiously. If I could even make it until 11:00 p.m., I'd need toothpicks to keep my eyelids open. I was diagnosed with clinical exhaustion and only had the bandwidth for work and sleep.

My energetic style was what contributed to my success at SoulCycle. Without it, would my classes sell out in minutes? Would there still be a waitlist of riders excited to get in? I knew I was on a path to burnout and depression. Along with the problem of simply not getting enough rest, the relationship tension was another huge energy suck. I was dragging ass when I needed to be kicking it, not only for myself, but for my students as well. Something had to change.

MICHAEL AGAIN

Stacey and I were each examining the crises in our own lives when we met for the first time in 2017, at an event in New York. We were both honorees for that year's Thrive Global Fuel List, me for my contribution to sleep education and Stacey as a fitness motivator. At the reception for all the honorees, a woman with platinum blond hair approached me and said excitedly, "I'm a *huge* fan of sleep and I've been stalking you for years!"

It was Stacey, of course. In her Googling for solutions to her sleep/relationship problem, she'd come across my TV appearances, books, website, and blog. If she was going to find a solution to her exhaustion, it would be through the Sleep Doctor. She asked if she could schedule an appointment with me to get to the bottom of her sleep issues.

We had a consultation and she shared her complaints. I realized immediately what was going on. I'm an expert in chronobiology (the study of circadian rhythms) and had written a book called *The Power of When* about the fact that not all people's genetically determined body clocks run on the same schedule. There are four different "chronotypes" and each one has a different inner schedule based on your genes. I'm a Wolf, a late riser who doesn't feel tired until midnight at the earliest. Stacey is an early riser, a Lion whose energy is highest in the morning and falls off a cliff after dinner. (These types used to be called "night owl" and "early bird," but I created a new classification system based on my research. The other two chronotypes in my system, along with Wolves and Lions, are Bears, who fall in between extremes, and Dolphins, the insomniacs.) Stacey's social life was more Wolfish, for those who can stay up late.

As I listened to her story, I knew she had to make some changes. Although I told her in a nicer way, I had to deliver a tough message: Lions can't function well if they try to live like Wolves. Stacey and her social life were incompatible. And her conflict couldn't be hacked because body clocks are genetic, part of one's DNA.

STACEY AGAIN

I heard Michael's underlying warning and knew it was the truth. Every day, every month or year, when I was fading, many of my friends would be emerging. We'd always be at our best at different times. As I knew from past relationships, there's something super special about waking up early together, having coffee or tea at the table as the sun comes up, preparing for sleep and getting in bed together at day's end. Shared rituals are intimacy enhancing. Incompatible chronobiology could widen a relationship gulf. It wasn't that I got tired early to be passive-aggressive or get out of going somewhere. I just needed rest and felt physically and emotionally awful if I lived out of sync with my body's natural rhythm.

The session with Michael was the first step in a major life change. But I received two parting gifts: (1) a much clearer idea of the characteristics to look for in close relationships, and (2) with Michael's help, a new understanding of myself.

The two of us — the Sleep Doctor and the Fitness Expert — started hanging out whenever we were in the same city, talking about our lives and updating each other on our vocations, namely, educating people about sleep and movement. During one meet-up in New York in 2019, we talked about the biggest complaint among his patients and my clients: "Why am I always so exhausted?"

This question was always topic number one whenever either of us did media interviews, too. It seemed like everyone was slogging through their days, sleeping fitfully all night, and living in a low-energy loop.

Michael had been studying and teaching the science of sleep for decades, and he knew that exhaustion was often the result of living out of sync with one's circadian rhythms, the inner clocks that dictated the ebb and flow of hormones, body temperature, and blood pressure. If you live *against* your natural rhythms—for example, your body wants you to go to sleep at 9:00 p.m., but you force yourself to stay up later night after night, as I had—you are at high risk for sleep deprivation,[1] chronic stress, mood disorders,[2] lowered immunity,[3] and compromised overall health. When you're out of sync with the master clocks in your brain (specifically, in your hypothalamus, I learned) and the dozens of mini-clocks that control every organ and system in the body, all your energy goes toward waging battles within. When you live in sync with your circadian clocks, you are not fighting against your own nature and have energy to spare.

OVER TO MICHAEL

In my reevaluation of my health after my trio of cardiac events, I discovered that I wasn't always following my own advice. Despite having literally written a book about how each of us needs to follow our own chronotype schedule for optimal health, I wasn't able to stay in sync with mine because of weekly travel to educate people about sleep! So I jammed my three weekly 5Ks to boost my energy—because everyone *knows* that exercise gives you energy and makes you healthy. But I didn't run during my chronotype's ideal exercise time (because I was usually on a plane), so I was not getting the benefits I thought I was—like stress reduction. As I learned in the scariest way possible, being healthy was not about checking one box, or doing what works for others. I needed to personalize an approach that took into account my chronotype as well as my eating schedule, anxiety level, and lifestyle.

Stacey suggested that I run *less* often. WTF?

Running three 5Ks a week—on top of working full-time (sixty-plus hours per week), traveling constantly, helping raise my kids—was more exhausting than energizing. Running caused stress instead of reducing it. It was just too much, especially for someone who was already under pressure, one step ahead of burnout, with a family history of heart disease. On top of that, because of relentless travel and constantly sleeping in different places, I had to deal with the first-night effect—having poor-quality sleep the first night in a new environment—which prevented my body from repairing and recovering from the stress I put myself under. It was a major "Doctor, heal thyself" moment and also a bit of a surprise. I ran a distance to reduce my stress, and when I had more stress, I just ran farther, because I had more stress, assuming that running more would make me *less* stressed. And now the fitness guru was telling me to exercise *less*?

Having seen the energy toll of overdoing it on some of her clients at SoulCycle, Stacey believes that more can be less, and less can be more when it comes to grueling fitness like running and spinning. Based on her observations, exhaustion was the result of the "more is always more" misconception. Exercise is important to stay limber, improve blood flow and oxygen circulation, and prevent muscle stiffness and fatigue that keeps people from wanting to move at all. Mathematician Isaac Newton's first law of motion (aka the law of inertia) states that an object in motion is more likely to stay in motion, and that an object at rest is more likely to stay at rest.

As Stacey explained, "Movement doesn't have to be constant, but it should be consistent."

She'd isolated *five particular times during the day* when people needed to get up and move to get their juices flowing: first thing in the morning, midmorning, after lunch, midevening, and before bed. A short movement break—taking a walk, doing twenty jumping jacks, stretching—didn't require the body to use up a significant amount of energy to signal the hormones that control alertness to "wake up."

Hearing Stacey's movement schedule got me thinking. The *same time periods* she mentioned coincided with hormonal changes in the Bear chronotype's circadian rhythms. (That's a lot of people! Bears account for half of the general population.) For example, the levels of cortisol, the fight-or-flight hormone, dropped in the midafternoon, around 2:00 p.m., which is why Bears feel sleepy after lunch. I often recommend a midafternoon "Nap A Latte"—drinking a cup of coffee, immediately taking a twenty-minute power nap, and waking up right when the caffeine kicks in. According to Stacey, doing three minutes of jumping jacks when you feel that afternoon dip in energy has the same heart-energizing, brain-clearing effects as a shot of espresso or a power nap. That gave me more options for my clients, which improved compliance.

Combining our ideas provided multiple recommendations for boosting energy!

STACEY AGAIN

We discussed the problem with one-size-fits-all energizing plans. Michael had his four chronotypes, and a different schedule for each. I pointed out that I recommended different exercise and movement approaches for the three main body types. You may have more of an understanding of body types than chronotypes since these were taught in high school health class. Do you remember endomorphs (curvy and rounded with a slow metabolism), ectomorphs (long and lean with a fast metabolism), and mesomorphs (athletic with a medium metabolism)? Our body type is determined by our genes. You can work with your DNA to become the best version of yourself, but you were born with your basic shape. **It turns out that each body type and its corresponding metabolic speed has a different capacity for effort, stamina, and power (all the aspects of movement).** Individuals respond to exercise or a single movement based on two important factors: their overall fitness history *and* their body type.

In my classes, I've noticed that naturally pear-shaped built-for-comfort people with a slow metabolism (endomorphs) tire quickly. An effective weight loss strategy for them is to do spurts of activity rather than maintaining a low, steady pace throughout class. For our purposes, let's call them Slow types.

Athletic, muscular people with a medium-speed metabolism (mesomorphs) gain energy when they push their limits, but if they work too hard, they can burn out or injure themselves (like Michael). Their best weight loss strategy is a combo of cardio and strength training. We'll call them Medium types.

Long-limbed and lean people with a fast metabolism (ectomorphs) have high endurance at a lower amplitude, but unless they build muscle, they won't feel as energetic as they could feel. They don't need a strategy to lose weight. Even if they put on extra pounds, they can lose the weight easily by eating a bit less and moving a bit more. They are Fast types.

As an instructor and personal trainer, I intuitively personalized workouts to fit clients based on their body type / metabolic speed so they'd have more fun and get better results. If clients were miserable at the end of a session, I knew they wouldn't keep up with the plan long-term. But for all types, I always emphasized that if you want to feel energized, you had to move your body at least five times per day, regardless of your exercise regimen, and of course get good sleep.

OUR STORY

Our discussion got even more fascinating when we correlated certain chronotype traits with body type. Early-riser Lions tend to be health-conscious, dedicated exercisers and have a lower-than-average body mass index (BMI). So are all Lions naturally athletic with a fast metabolism? Probably not all.

Late-riser Wolves are less health conscious with higher BMIs. Does that mean a typical Wolf is a built-for-comfort, slow-metabolism endomorph? Most likely not.

Dolphins (insomniacs) tend to be naturally slim and indifferent to fitness routines, so are a significant majority of them the long-and-lean ectomorph type? Doubtful.

Bears, or "in the middle" people, have average to above-average BMIs and care about their health theoretically but aren't obsessive about fitness or healthy eating. Is it possible that most of them are comfy, slow-metabolism endomorphs, too? Remained to be seen.

We started to develop our own theory. But to prove it, we had to do some scientific research and conduct studies of our own to get statistics on these cross-referenced chronotypes and body/speed types. With that information, we could create *personalized sleeping and movement plans for every possible combination*. What would be the ideal schedule and the ideal movements for a Wolf with a medium metabolism like Michael, or a Lion with a fast metabolism like Stacey?

We knew that being sedentary was tiring and that moving was energizing. But how could we get driven Lions and neurotic Dolphins to plan more rest and recovery in their day so that they could prevent burnout? And what about Wolves, the rebels, the night creatures who hate exercise and healthy food and would rather have a beer than attend a fitness class? What kind of daily movement plan would entice a slow-metabolism Wolf to get off the couch?

Any energizing protocol had to be highly individualized. One-size-fits-all plans were useless. Everyone is preprogrammed with DNA for a specific chronotype and metabolic type. You were born to be a Slow Bear or a Fast Lion; it's in your genes. So stop wasting energy by living out of sync with who you are. *Instead, use the simple science of good timing and small movements to reduce energy expenditure and increase energy replenishment.* If people used their own DNA to their advantage, they'd get more benefit from movement and would improve the quality of their sleep to sufficiently recharge and power up their bodies for maximum energy.

To stop feeling "so tired" and burst with infinite energy all day long, all you have to do is follow the right schedule, personalized for your chronotype and body type.

Then we just had to figure out what the right schedule was for each type.

The good news: We did it. We worked our asses off compiling data, including from the results of a chronotype/body-type quiz that we sent out to Michael's online community. Five thousand people took it, giving us deeper insight into the combinations and types of people out there. With the data, we isolated eight distinct **Power Profiles** and calculated eight corresponding **Power Protocols,** with a daily schedule for five timed movement sessions (each about five minutes long). But we didn't stop there. Since people gain energy by eating and sleeping on the right schedule, we figured out the ideal schedule for those activities, too. And we used scientific research to identify specific energy drains for each type and came up with strategies for combatting them.

The Power Protocols have been road tested by members of our communities with amazing results. Weight loss. Stamina. Less stress. Better endurance and strength. Brightened mood. More satisfying relationships. Higher-quality sleep.

Our two areas of expertise—circadian timing and movement—truly came together to create a whole new concept of how people can improve their health and feel more of what we all desperately crave: energy. Going from "I'm so *tired!*" to "Hell *yes!*" is a simple matter of knowing when to move, how to move, when to rest, when to eat, when to stop eating, and how to boost your mood.

Tweak your schedule, increase your movement, supercharge your energy, change your life.

TIMING + MOVEMENT = ENERGY

This formula works. It's the antidote to exhaustion, the secret to accessing your abundant natural resources, to setting off sparks in every area of your life. Our plan is so successful because it asks you to be authentically yourself.

We're so psyched to share this with you! It's going to change your life.

Personal experience taught us that what works for one person does not work for all. Running three 5Ks won't necessarily boost energy and reduce stress. Staying up much later than your genetic bedtime can't be brushed off as no big deal. But personalization allows the body to work at its peak potential. Science is finally teaching us that people are unique, and that if we try to sleep, move, or eat on someone else's schedule, we are making ourselves tired and vulnerable.

Working together, we cracked the code for how you, your mother, your friend, your partner, your kids, or your boss can achieve personalized, energized greatness.

We know there are ideal times to sleep based on our chronotype.

We know there are ideal ways to move based on our body type.

With a personalized plan for beating exhaustion with timing, movement, eating, and sleep schedules, you can eradicate exhaustion entirely. No dieting, no deprivation, no hard-core exercise. By following your personalized protocol, you tap into the vast power and potential of "doing you," genetically and energetically speaking. We're opening our toolbox here, and you'll have all you need to biohack your energy reserves, end exhaustion, and put yourself on a faster track for success and wellness.

UNPACKING EXHAUSTION AND ENERGY

CHAPTER I

Exhausted to Energized: The Energy Scale

What's holding you back from making positive change? Exhaustion and doubt. By definition, exhaustion is being really drained, feeling spent, sucked dry, like a hollow shell of the person you used to be and hope to be again. The unscientific term we use that seems to cover the essence of exhaustion is "dragging ass."

Our patients and clients tell us:

"I want to feel more alive."

"I want to feel more awake."

"I want to be alert all day and have a restful night."

"I'm doing everything I'm supposed to do, but I'm still so tired."

Are *you* stuck in your exhaustion? Are you too tired to get everything done so you push things off till tomorrow, locked in a forever game of catch-up? As a nation, we're overwhelmed and overloaded, comforting ourselves with Netflix and carbs; procrastinating feels like it's our second job.

Finding the energy and motivation to break this cycle might seem like the impossible dream for now. But with the science explained in this book, *you will turn exhausted into energized* and become *that* person, the one with the on switch that just doesn't quit. All you need to live a full life — with satisfaction and confidence to spare — is to know your genetic code and be the person you were born to be.

> ⚡ Energize Tip: When you sleep, move, and eat on a schedule that personally fits *your* genes — not someone else's — energizing is easy!

Before you pick up energizing tools and start swinging them, first we need to unpack the basics about what energy is and why it's the one thing we all need more of to be the productive, dynamic, effective, self-motivated humans we all long to be. There are several definitions of energy. The simplest scientific definition:

ENERGY IS "THE ABILITY TO WORK"

Work isn't just meeting your professional responsibilities, although that is part of it. The ability to work means having the power to *go somewhere* and *do something*. If you have enough gusto to do most of what you need to do and get where you need to go, you are energized enough to function adequately. Functioning a bit better than "adequately" is so easily achieved, it would be tragic not to aim higher.

Many of us can't remember a time when we felt full of life. We've become accustomed to functioning just well enough to complete our to-do lists and hit our deadlines. When every day feels like a slog, your body's internal energy "body battery" is only partially charged. If your battery were four-bars fully charged—which the strategies in this book will guide you to do very quickly—life would feel different. You'll be able to do everything you have to do and have tons of juice left over for the things you want to do. It will feel like you've finally found that lightning in a bottle. Ironically, it's always been inside you. Most likely, it's been dimmed by bad timing, poor sleep, and not enough movement. All that's going to change, quickly.

Honestly, chores will always be tedious. Deadlines will be stressful. Conflicts will take a lot out of you. The **Power Protocols** in this book won't change that. But the demands of life that seem exhausting to you *now* won't send you back to bed or straight to the couch anymore. If you make small tweaks to your schedule, what used to deplete your energy will barely make a dent.

HOW EXHAUSTED ARE YOU?

How exhausted we think we are might not match up with how exhausted we are on a purely physical level. It's a matter of perception.

One way to sort this out is an assessment tool originally developed by Swedish psychologist Gunnar Borg.[1] The "rate of perceived exertion" (RPE) measures how hard you think you're working when exercising. An RPE of zero indicates that you are at rest. An RPE of ten means that you are going full out, maximum effort and intensity, working as hard as you have ever worked.

THE BORG RATE OF PERCEIVED EXERTION SCALE

Level	How It Feels
Level 10: Maximum Activity	Out of breath, unable to talk, feel like you can't go on
Level 9: Very Hard Activity	Difficult to breathe, can only say a few words at a time, can barely maintain it
Levels 7 and 8: Vigorous Activity	Borderline uncomfortable, short of breath, can speak a sentence at a time, it's hard to keep going
Levels 4, 5, and 6: Moderate Activity	Breathing heavy, can speak a paragraph at a time, exercise is challenging but sustainable
Levels 2 and 3: Light Activity	Breathing easy, can carry on a normal conversation, exercise sustainable for hours
Level 1: Very Light Activity	Barely moving at all

The Borg scale is a useful tool for athletes to assess their effort. But perceived exertion can be extrapolated to how we feel just getting through the day. It's not that you aren't working hard or taking things easy. It's about how you perceive your effort. Small tasks like walking the dog or taking out the garbage—objectively light-effort activities—might *seem* like level 8 undertakings if you are chronically exhausted.

When you adjust your schedule and commit to timed movements and specific sleep hours, your body battery will stay fully charged, and the busywork of life won't seem all-consuming. Otherwise-draining hardships won't send you spiraling downward or kill your capacity for fun. Your "ability to work," work out, work things out, and get shit done will launch you into untapped success, and it will feel almost effortless.

Another scientific definition of energy:

ENERGY IS "THE ABILITY TO CHANGE"

An external energy source can trigger internal change. For example, a plant changes sunlight (external energy) into growth (internal energy). Animals change food energy into kinetic energy to walk, swim, and fly. Movement energy, like strumming a guitar, can turn into sound energy. If you look at energy as a way to change one form into another, energy is pure potential.

> ⚡ Energize Tip: In terms of our personal human potential, energy is not only the ability to change, but the essential ingredient we need to make change possible.

If we're too strung out mentally and physically to take on new challenges, we'll never be able to make the changes we fantasize about. Whether it's "get in shape," "fall in love," or "make more money," change depends on having the energy to go for the win, not the tie.

YOUR ENERGY GOALS

Your first steps toward realizing your energy goals are to identify them and write down a plan of action.

Michael's energy goals:

1. Daily meditation: At 7:00 a.m., he will use his Muse meditation headband or BrainTap meditation tracker for fifteen minutes.
2. Daily breathwork: At 7:25 a.m., he will follow an online video or join a group of friends in METAL International to do breathwork for fifteen minutes.
3. Daily exercise: At 8:00 a.m., he will do group or individual fitness for half an hour.

Stacey's energy goals:

1. Sleep: Get seven and a half hours per night by setting a bedtime and unplugging an hour prior.
2. Exercise: Do one to two hours per day through instruction or on her own.
3. Intermittent fasting: Fast for fifteen hours per day by using a countdown app like Zero or doing it with a friend.

List three positive changes you'd like to make if you had more energy and what you will do to make them REAL!

1. Goal:_____ Plan:_____

2. Goal:_____ Plan:_____

3. Goal:_____ Plan:_____

MICHAEL SAYS...

I have patients stuck in the mental-exhaustion washing machine. They talk about their horrible bosses and ruminate about work stress long into the night. (I call it giving rent-free space in your brain!) Their job anxiety robs them of the quality sleep they need to recharge their battery so they can wake up with the energy to face off with their horrible boss again. When I suggest they might consider looking for a new job that isn't as anxiety provoking and sleep disrupting, they often say, "Just the idea of posting my résumé *exhausts me*." These patients fully acknowledge that their circumstance has sucked their energy dry, but they are too drained to do anything about it. Unhappy job, toxic relationship, unhealthy lifestyle...these are Energy Vampires. They deprive people of the energy they need to lift themselves out of a bad situation.

STACEY SAYS...

And I have clients who do two spin classes in the morning, laboring under the idea that intense exercise will boost their energy. But they are so exhausted afterward that they have to push themselves to get through the rest of their day. They drain their reserves at a rapid rate without allowing their bodies to recharge. When I tell them, "You'll feel better if you do less," they don't believe it. All their lives, they've been told that exercise increases energy. So the more they exercise, the more energy they should have. It's just not true. Remember Michael?

As a nation, we are *wrung out*. According to a recent survey[2] of 1,011 American adults by the National Sleep Foundation:

- Half of the respondents said they feel excessively tired three to seven days per week.
- Exhaustion negatively affects their mood, focus, motivation to exercise, and ability to "get things done."
- It makes them feel irritable and impairs their willingness to go out and have fun in the evening.
- They report having headaches and general "unwell" symptoms.
- A third of respondents said that tiredness takes a hit on their work performance.
- A quarter of respondents said exhaustion affects their personal relationships.

You won't be surprised to hear that stress and sleepiness (and their adverse effects on mood, performance, well-being, relationships, health, and potential) are related. More stress = more sleepiness = lower energy.

We need to clarify the distinction between chronic exhaustion and chronic fatigue. Chronic exhaustion is feeling sleepy at least two or more days per week due to lifestyle choices and conditions like insomnia and anxiety. It's different from myalgic encephalomyelitis, aka chronic fatigue syndrome, a systemic disease. Symptoms of ME/CFS include muscle and joint pain, muscle weakness, sore throat, whole-body fatigue for six months that gets worse with exertion, cognitive fog, lack of concentration, excessive sleep, increased anxiety, and depression. If you have these symptoms, please go to your doctor for a diagnosis and treatment.

Another key finding of the survey: Despite their full awareness that exhaustion was harming their lives in just about every area, two-thirds of the respondents said they believe they can just "shake it off and keep going," using diet soda, crappy snacks, and so-called energy drinks as their primary coping strategy.

> ⚡ Energize Tip: Empty-Battery Zombies wake up tired, jack up on coffee and doughnuts, drag ass all day, and combat fatigue with more sugar and more caffeine, stimulants that make it harder to get adequate quality sleep...and repeat the exhaustion cycle for eternity.

Along with feeling low and slow, we're plagued by an insidious side effect of exhaustion: stagnation. We get stuck in vicious cycles that make any undertaking feel too big and way too tiring to attempt. A week or a year or a decade goes by in the exhaustion-stagnation loop, and we don't make the changes we want for ourselves. A once-a-year weekend retreat is not going to cure this. But a life shift can.

Are you too tired to...

Date?

Hang with friends?

Exercise?

Make it rain at work?

Cook your own meals and eat well?

Be active with your kids?

Apply to grad school?

Look for a new job?

Have sex with your partner?

Write that screenplay?

> ⚡ Energize Tip: With more energy, you can reimagine what you can accomplish in your life and make that vision a reality.

If so, energy is the factor that will decide your future health, happiness, and level of success. By recalibrating your life to increase your energy, you can do a complete one-eighty. With energy, you can redefine who you are and what you have yet to accomplish. What's holding you back? Nothing! Now you have *us*. Sometimes you simply need the right coaches to guide you, and that's what we plan to do.

ENERGY GAINS VS. ENERGY DRAINS

Our bodies are amazingly complex, but our energy expenditure and replenishment system is pretty basic. As long as your energy gains are equal to your energy drains, you'll have what you need to function on a basic level. If your gains *exceed* your drains, you can function at a super-high level, which is what we want. To do that, you have to know what drags you down and what lifts you up, and then aspire to have more checks in the *up* column.

By following your personal Power Protocol, you'll build gains and minimize drains as your sleeping, moving, and eating schedules turn your body into a super-functioning recharge station.

↑ **Energy Gains**

- Adequate high-quality sleep
- Power naps
- Moderate-intensity movement
- Post-fitness rest and recovery
- A healthy weight
- A balanced diet
- Hydration
- Meditation
- Planned mealtimes

- Time outside
- Intimate relationships
- Community involvement
- Music
- Laughter
- Fun
- Good timing!

These gains aren't one-size-fits-all. What gives *you* a boost might be draining for someone else. A Lion (like Stacey) would benefit from a power nap every day for a quick gain. But if a Dolphin took a long nap in the middle of the day, she'd get lost in thick brain fog all afternoon and wouldn't be able to fall asleep at night, setting up fatigue for the next day. For some, going to a big party full of strangers is superexciting and floods their battery with power. For those with social anxiety, the same situation might be instantly depleting.

Based on the science and our experience working with thousands of people, we've tailored specific energy goalposts for each Power Profile, and we describe how you can get there later in the book. Part of our strategy is for each of you to identify activities that, for you, are like plugging your power cord into the sun.

The other half of the energy equation is the minuses, the drains, the stuff you want to eliminate from your life as much as possible.

↓ **Energy Drains**

- Inadequate, poor-quality sleep
- Too much caffeine
- Too much sugar
- Too much ass-in-chair time
- Oversleeping
- Overexercising
- Drama
- Emotional negativity
- Excess pounds

- Dehydration
- Snacking after dinner
- Drugs
- Alcohol
- Isolation
- Physical and emotional stress
- Anxiety
- Depression
- Bad timing!

We do *not* want to make anyone feel bad about their weight. There's way too much body shaming in the world already! We're all about embracing the body you were born with and working with it to achieve your fitness and mental health goals. One of those goals might be to free yourself of excess pounds, which, according to the research, do drag energy down.

In a recent study, researchers at Texas A&M University and the University at Buffalo examined how weight affected "fatigability."[3] They asked their participants—49 normal-weight, 50 overweight, and 43 obese adults—to do tiring exercises at a range of exertion levels. The obese participants had up to 30 percent lower endurance than their normal-weight counterparts and a "faster progression in perception of effort," meaning they thought they were working harder and quicker than the other participants.

Weight can be a huge factor in rest-disrupting (and energy-depleting) sleep apnea. A high-sugar diet and/or late-night pig-outs, and the resulting excess pounds, cause stress on the body via hormonal havoc and inflammation, which also affect sleep quality. Research has found that it's extremely difficult to lose weight if you don't sleep well.[4] (Read Michael's book about this phenomenon, *The Sleep Doctor's Diet Plan: Lose Weight Through Better Sleep*.) If you do care about dropping pounds, our strategies will help you do it, and keep doing it, with a TikTok dance in your step and an Instagram selfie-ready smile on your face.

Drugs and alcohol disrupt sleep and affect your motivation to go outside and soak up some energy-infusing vitamin D. Alcohol is dehydrating, and being dehydrated has been proven to make you feel tired and sluggish,[5] like an Empty-Battery Zombie with a hangover, for a day or two after.

> ⚡ Energize Tip: Nearly everything you do and feel has an impact on your energy. You have the power to nudge your energy in the right (or wrong) direction by how you behave. *You are in control of your energy.*

WHAT'S YOUR ENERGY SCORE?

Before we get into the science of how a revamped schedule will drastically increase your energy levels and make everything seem a lot easier, we have to assess your current energy levels. You need a personal baseline against which to measure your future leaps and bounds.

To know where you're going, you have to know where you are.

Along with "the ability to work and change," energy is, as we all intuitively understand, a *feeling*. How you perceive your exertion affects how you *feel* about what you can and can't do. How you *feel* from hour to hour determines how well you can work and change for the better. Every hour matters. At any given point during the day, you might feel up (wide-awake, ready to *go* and *do,* happy, psyched) or down (sluggish, ready to drop your head on your desk, miserable, unmotivated).

You might be thinking about taking a walk, tackling a simple task, painting your house, or having a difficult but necessary conversation. If you self-scan and decide, *Nope. Not going to happen today,* that's another twenty-four-hour period of your life when low energy interfered with your sense of accomplishment and satisfaction.

When you focus your mind on one question— "How do I *feel* right now, energywise?"—you can begin to see how your perceived power level affects nearly every decision you make throughout the day, in your relationships, career, and physical and mental health.

By gathering stats about how you feel energetically at five key times throughout the day, you can calculate a baseline measurement to build on.

When we began testing people on our protocols, most of them tended to overstate their baseline energy levels, only because they didn't fully grasp just how tired they were to begin with. Feeling exhausted was a chronic state; it felt normal. After our road testers followed their Power Protocols for several weeks, they slept better, ate on a fixed schedule, and moved more throughout the day. They didn't realize how low and slow they had felt until they started turning the energy dial to eleven.

To track your own energy level, over the course of a week, you'll fill in the Energy Diary later in this chapter at the five key times outlined in the following list using a scale of 1 (lowest) to 10 (highest). Seven days of data will provide big clues about how your energy level rises and falls over the course of the day and about how you feel overall. Set alarms on your phone and check in at these times:

- Upon waking
- Midmorning (three or four hours after waking)
- Midafternoon (one to two hours after lunch)
- Evening (around dinnertime)
- Right before you get in bed at night

HOW TO RATE YOUR ENERGY LEVEL FOR THE FIRST FOUR CHECK-IN TIMES

- **Level 1: Very low energy.** Yawning very frequently (every few minutes), reaching for caffeine and sugar in order to stay awake, extreme difficulty keeping eyes open, slumped posture, moving feels impossible, forced to say "no" to potentially fun activities.
- **Level 2 to 3: Low energy.** Yawning frequently (every ten minutes), reaching for caffeine and sugar to stay alert, feeling like you could fall asleep at any moment, slumped posture, moving feels painful, still have to say "no" to potentially fun activities.

- **Level 4 to 5: Moderate energy.** Yawning triggered by others doing it, reaching for caffeine as a pick-me-up, awake but not completely alert, stooped posture, moving feels doable, have to say "probably not" to potentially fun activities.
- **Level 6 to 7: Adequate energy.** No yawning, reaching for caffeine for pleasure, awake and alert, upright posture, moving feels fine, say "okay" to potentially fun activities.
- **Level 8 to 9: High energy.** Yawning unthinkable, no need for caffeine, wide-awake and alert, straight posture, moving feels enjoyable, say "yes" to potentially fun activities.
- **Level 10: Maximum energy.** Yawning unthinkable, caffeine uncomfortable, eyes bright, brimming with alertness, running on eight or more hours of sleep, perky and dynamic posture, moving feels fabulous, say "hell yes!" to potentially fun activities.

At bedtime, your energy level is in reverse. If you are appropriately tired and able to fall asleep quickly, you get a higher score because getting a good night's rest is energizing. You get a lower score if you are wide-awake and unable to sleep because sleep disruption is exhausting.

HOW TO RATE YOUR ENERGY LEVEL AT BEDTIME . . . ZZZ

- **Level 1 to 2: Wide-awake.** There is no way possible you could sleep now.
- **Level 3 to 4: Awake.** You aren't anywhere close to feeling sleepy; it'll be hours before you are.
- **Level 5 to 6: Tired and wired.** You're dragging, but you're too full of nervous energy to sleep yet.
- **Level 7 to 8: Tired.** You're yawning and your eyelids feel heavy. Bed is calling.
- **Level 9 to 10: Very tired.** You know that you will pass out within minutes of your head hitting the pillow.

SAMPLE ENERGY DIARY

	Monday	Tuesday	Wednesday	Thursday	Friday	Saturday	Sunday
Waking	2	2	3	4	3	2	4
Midmorning	5	6	6	5	6	7	7
After lunch	2	3	3	4	4	5	3
Evening	6	7	6	5	7	7	7
Bedtime	4	3	5	5	5	4	5
Daily Average:	3.8	4.2	4.6	4.6	5	5	5.2
Weekly Average:	4.6						

The point of keeping an Energy Diary *before* you begin our program is to get to know yourself and your natural energy ebbs and flows. It's normal to feel more power on some days than others (women, see the box on page 297). For example, Monday might be a low-energy day for you for various reasons (which we'll get to later). If you have a poor night's sleep, your next day will have a lower baseline. Just get familiar with your energy and begin to notice how and why it fluctuates. Assessing how you feel first and last thing every day will keep you motivated to make positive changes. You'll continue to track your energy as you go through our month-long program, and you'll have proof of how it's working.

Both of us looked back at our Energy Diaries prior to starting the program, and here is what we saw:

MICHAEL'S ENERGY DIARY

	Monday	Tuesday	Wednesday	Thursday	Friday	Saturday	Sunday
Waking	2	2	2	2	3	2	2
Midmorning	4	4	5	5	5	4	5
After lunch	3	3	3	3	3	3	3
Evening	6	6	6	7	7	7	6
Bedtime	5	5	6	6	6	5	5
Daily Average:	4	4	4.4	4.6	4.8	4.2	4.2
Weekly Average:	4.3						

This was a low-energy week for Michael. He was traveling and under stress and it showed. Typically, as a Wolf, he is tired upon waking, has cleared his morning fog by midmorning, and has a slump in the early afternoon. By evening, though, he hits his energy peak, which continues until bedtime. As you can see, he's often tired and wired at bedtime and not ready to go to sleep anytime soon.

STACEY'S ENERGY DIARY

	Monday	Tuesday	Wednesday	Thursday	Friday	Saturday	Sunday
Waking	9	9	9	8	9	8	9
Midmorning	8	8	8	7	8	8	7
After lunch	7	7	7	8	8	8	7
Evening	5	5	5	6	5	5	6
Bedtime	9	9	9	9	9	9	9
Daily Average:	**7.6**	**7.6**	**7.6**	**7.6**	**7.8**	**7.6**	**7.6**
Weekly Average:	**7.6**						

In Stacey's typical week, she wakes with peak energy in the morning and rides that wave all the way through into the afternoon (boosted by an energy-gaining nap). But her Lion power starts to wane in the evening and crashes completely at bedtime, when she can barely keep her eyes open.

Everyone is unique, and we can't promise that each person who starts with an average weekly score of 3 or 4 will double it in one month's time. But if you make a commitment to follow the plan, we know you will make huge energy gains. Imagine how your life would change with even a little more energy than you have now. Amazing!

With good timing and movement, you won't have to imagine it. You'll be living it.

YOUR ENERGY DIARY

	Monday	Tuesday	Wednesday	Thursday	Friday	Saturday	Sunday
Waking							
Midmorning							
After lunch							
Evening							
Bedtime							
Daily Average:							
Weekly Average:							

TAKEAWAYS: EXHAUSTED TO ENERGIZED

- Empty-Battery Zombies survive on caffeine, but living with chronic exhaustion is taking a huge toll on them, physically and emotionally.
- Clue in to your energy by asking, "How do I *feel* right now, energy-wise?" to connect the dots between your feelings and your decisions.
- Identify energy gains that, for you, are like plugging your power cord into the sun.
- Identify energy drains, the stuff you want to eliminate from your life as much as possible.

TO DO RIGHT NOW!

1. List your energy goals. Write them down.
2. Set alarms on your phone to check your energy levels. It's easy.
3. Fill in a week of data in your Energy Diary *before* you start the program. You've got to know where you start to see the finish line.

CHAPTER 2

What's Your Power Profile?
Body Type and Chronotype

WHAT IS A BODY TYPE?

Just as people are born with genes for brown or blue eyes, we are all born to fall into one of three body types:[1]

1. Ectomorphs (lanky with a fast metabolism)
2. Mesomorphs (muscular with a medium metabolism)
3. Endomorphs (curvy with a slow metabolism)

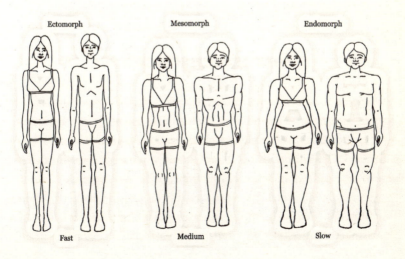

> ### STACEY SAYS...
>
> As a trainer and instructor for decades, I've guided tens of thousands of clients of all sizes and genders into the best shape of their lives. And I've learned that what works for one person does not necessarily work for another. What matters is fitness level, which is in flux, and body type, which does not change, no matter how ripped you get. I believe that the number one cause of exhaustion is people striving to become a body type that they were not born to be, and all the soul- and energy-crushing self-hate people come to feel about their bodies when they have a certain genetic shape.

↓ **Energy Drain: Body-Expectation Mismatch**

↑ **Energy Gain: Self-Acceptance**

The most important source of energy is being active in general and doing the five daily timed movement sessions that respond to what *your* body needs *at that time*. Once you understand and appreciate how your body functions best, you'll move and metabolize most effectively and energetically for you.

Half of your Power Profile is your body type and its metabolic speed.

WHAT IS A CHRONOTYPE?

Inside your brain and body, there are dozens of clocks that control every organ and system. They tell you when it's time to sleep, eat, think, move, rest, and digest. Not all biological clocks run on the same time. People are different! Determined by our genes (thank your parents), we are all born to be one of four different chronotypes, with corresponding chrono-rhythms, or bio-schedules—early risers (Lions), in-betweens (Bears), night people (Wolves), and insomniacs (Dolphins). Scientific research has linked chronotype to sleep habits as well as personality traits,[2] behaviors,[3] cognitive ability,[4] and outlook on life.[5] Your body's particular schedule reveals a lot of information about who you are.

> ### MICHAEL SAYS...
>
> The number one cause of exhaustion is people not following their genetic chronotype's timing. They might be trying to stick with a societal schedule that doesn't line up with their internal timing. Or they might have a partner whose bio-timing is a mismatch with their own. Remember Stacey's situation? Energy drain is due to bad bio-timing; the most important energy gain is to stick to good timing.

When you live according to your chronotype, you are in sync with the schedule your body was born to follow. By not fighting that timing, you sleep and digest better and free up energy for sharper focus and motivation. Once you know your chronotype and adjust your daily schedule and lifestyle to fit your natural-born chronorhythm, every system and organ in your body will run more efficiently and *energetically*.

Half of your Power Profile is your chronotype.

Your Power Profile—metabolic speed + chronotype—gives a detailed portrait of your body, your habits, and your personality. Once you know your profile, you can reorganize your life according to your Power Protocol, a daily schedule that will provide maximum energy for *you*. It's your map to a speedier metabolism, better sleep, and a more sexy, strong, confident you. You could call your Power Protocol a magic formula, but it's based on pure biology.

You might have an idea about your chronotype and body type already, but let's take a dive into the science.

WHAT'S YOUR BODY TYPE?

To figure out which body type you are, complete the following quiz as honestly as possible:

1. **When you were nine or ten, you were:**
   a. Skinny
   b. Average weight
   c. Stocky

2. **If you wanted to drop a few pounds, it would be:**
   a. Easy!
   b. Doable with effort
   c. Extremely difficult

3. **If you wanted to gain a few pounds, it would be:**
   a. Tough — it's hard enough to keep weight on, let alone gain it
   b. Easy — fewer workouts and more pasta should do it
   c. Piece of cake! In fact, just looking at the cake...

4. **Which is wider, your shoulders or hips?**
   a. They're the same
   b. Shoulders
   c. Hips

5. **Choose the shape that best describes you:**
   a. Straight line
   b. Hourglass (for women); V-shaped (for men)
   c. Bottom heavy (for women); thick around the middle (for men)

6. **Try to touch your middle finger and thumb around your opposite wrist.**
   a. They overlap
   b. They just touch
   c. Can't quite do it

7. **You are hungry:**
   a. Rarely
   b. Just at mealtimes
   c. All the time

8. **During long meetings at work:**
   a. After half an hour, you start to fidget
   b. After an hour, you feel a bit restless
   c. You could sit still all day

SCORING

Add up the number of (a), (b), and (c) responses. Which letter did you circle the most?

Mostly (a): Ectomorph

Mostly (b): Mesomorph

Mostly (c): Endomorph

No matter what, you are awesome!

Ectomorphs

Ectomorphs (fast-metabolism types) are born to be long and lean, with graceful limbs and swan-like necks. They won the genetic lottery to be supermodels in the '80s, but they're not guaranteed to have the ideal body forever. We all know some lanky people who carry a bit of extra weight around the core. Some characteristics:

- Naturally skinny, often tall
- Slim, long limbs
- In constant fidgeting motion
- Rarely hungry; eat to live (and yet some are foodies)
- Highly effective at using calories and stored fat for energy
- Ineffective at building muscle
- Weight loss is easy
- Prefer non-team fitness
- Excellent endurance athletes (think marathoners)
- Not predisposed to any diseases or obesity

FAMOUS ECTOMORPHS

These peeps have a fast metabolism:

- Gwyneth Paltrow, Zoe Saldana, Nicole Kidman, Misty May-Treanor, and Maria Sharapova. For women, it's all about spidery long arms and legs, and not much booty.
- Brad Pitt, Zac Efron, Ryan Gosling, Neil Patrick Harris, Usain Bolt, Michael Phelps, and Bruce Lee. The giveaway for men: No matter how much bulking up they do at the gym, they still have small wrists.

Mesomorphs

Mesomorphs (medium-metabolism types) are muscular and athletic, people who look like they could up and sprint a mile or do back handsprings on demand. They have visible abs without trying. But that doesn't mean they'll stay at their college weight and shape forever. Some characteristics:

- V-shaped male body; hourglass-shaped female body
- Shoulders equal to or wider than hips
- Athletic
- Hungry at mealtimes
- Effective at burning calories and stored fat for energy
- Highly effective at building muscle
- Effective at storing fat
- If they don't maintain their muscle mass, they will lose it eventually
- Prefer team sports and inspired by competition
- Excellent power athletes
- At risk for digestive-system diseases, hypertension, liver disease[6]

FAMOUS MESOMORPHS

These peeps have a medium metabolism:

- Halle Berry, Gal Gadot, Emma Stone, Megan Rapinoe, and Ronda Rousey. For women, think hourglass shape with shoulders as wide as (or wider than) hips.
- David Burtka, Mark Wahlberg, Cristiano Ronaldo, and Will Smith. The giveaway for men is their V-shaped physique and medium-size wrists.

Endomorphs

Endomorphs (slow-metabolism types) tend to be curvier; they carry more weight around their hips, butt, and belly. They need to be careful about weight gain through exercise and nutrition, but endos have the body type that is popular in our culture right now. Some characteristics:

- Round male body; pear-shaped female body
- Narrow shoulders, wide hips, shorter limbs
- Less likely to participate in sports or fitness activities
- Hungry all the time
- Ineffective at burning calories and stored fat for energy
- Moderately effective at building muscle
- Highly effective at storing fat
- If they don't maintain their muscle mass, they will lose it quickly
- Prefer low-impact or no exercise; easily frustrated when exercise doesn't "work"
- High potential to become athletic
- High risk for digestive-system diseases, hypertension, liver disease, obesity[7]

FAMOUS ENDOMORPHS

These peeps have a slow metabolism:

- Jennifer Lopez, Beyoncé, Cardi B, and Serena Williams. Female endomorphs have shorter limbs, generous behinds, smaller shoulders, and thicker wrists.
- Chris Pratt, Chris Hemsworth, Tom Hardy, Jack Black, and every linebacker in football. The giveaway for men is shorter limbs, thicker middle sections, and overall stockiness.

No amount of SoulCycle classes or Rumble workouts is going to turn a curvy endomorph into a lanky ectomorph. Ice cream on tap will not turn a muscular mesomorph into a voluptuous endomorph. You'll still be you, but with another 10 pounds on your genetically determined shape.

If you try to become someone you're not, you will only experience frustration and disappointment. But by accepting your type, you can become the strongest, healthiest, hottest version of who you already are and enjoy the physical and emotional energetic benefits that come from being in sync with yourself.

If you aren't sure of your type or think of yourself as falling between types, that's normal. Hardly anyone is 100 percent just one. In fact, humans are trending toward being endomorphic,[8] and most of us are a combo of ecto/meso and endo/meso.

My body type [ectomorph, mesomorph, or endomorph] is

_____.

My metabolic speed [fast, medium, or slow] is

_____.

WHAT'S YOUR CHRONOTYPE?

Chronotype is not just your genetic predisposition to wake early or stay up late. It's a classification system that includes far more about who you are as a person, how your mind works, your ambition, and your lifestyle choices, like diet and exercise. There truly is a best time to do just about everything, as long as you know your type.

You might have heard of three of the four types: early bird (early riser), night owl (stays up late), and hummingbird (in between). Michael reinvented a classification system that better reflected his research: Lions (early risers), Bears (in-between types), Wolves (night people), and Dolphins (insomniacs who are always half-awake).

To find out *your* chronotype, take the Chronotype Quiz online at Chronoquiz.com or complete the two-part quiz right here:

Part One

For the following ten statements, please circle T for true or F for false.

1. **The slightest sound or light can keep me awake or wake me up.**
   T or F
2. **Food is not a great passion for me.**
   T or F
3. **I usually wake up before my alarm rings.**
   T or F
4. **I can't sleep well on planes, even with an eye mask and earplugs.**
   T or F
5. **I'm often irritable due to fatigue.**
   T or F
6. **I worry inordinately about small details.**
   T or F

7. **I have been diagnosed by a doctor or self-diagnosed as an insomniac.**
   T or F

8. **In school, I was anxious about my grades.**
   T or F

9. **I lose sleep ruminating about what happened in the past and what might happen in the future.**
   T or F

10. **I'm a perfectionist.**
    T or F

If you marked T for true on **seven or more** of the above ten questions, **you are a Dolphin** and can skip ahead to the "Dolphins" section later in this chapter.

Otherwise, continue on to . . .

Part Two

After each of the multiple-choice options, you'll find a number in parentheses. Keep a tally of these numbers to get your final score.

1. **If you had nothing to do the next day and gave yourself permission to sleep in as long as you like, when would you wake up?**
   a   Before 6:30 a.m. (1)
   b.   Between 6:30 a.m. and 8:45 a.m. (2)
   c.   After 8:45 a.m. (3)

2. **When you have to get out of bed by a certain time, do you use an alarm clock?**
   a.   No need. You wake up on your own at just the right time. (1)
   b.   Yes to the alarm, plus one or two snoozes. (2)
   c.   Yes to the alarm, with a backup alarm and multiple snoozes. (3)

3. **When do you wake up on the weekends?**
   a.   The same time as your workweek schedule. (1)

    b.   Within forty-five to ninety minutes of your workweek schedule. (2)

    c.   Ninety minutes past your workweek schedule. (3)

4. **How do you experience jet lag?**

    a.   You struggle with it, no matter what. (1)

    b.   You adjust within forty-eight hours. (2)

    c.   You adjust quickly, especially when traveling west. (3)

5. **What's your favorite meal? (Think time of day more than the menu.)**

    a.   Breakfast. (1)

    b.   Lunch. (2)

    c.   Dinner. (3)

6. **If you were to flashback to high school and take the SAT again, when would you prefer to *start* the test for maximum focus and concentration (not just to get it over with)?**

    a.   Early morning. (1)

    b.   Early afternoon. (2)

    c.   Midafternoon. (3)

7. **If you could choose any time of day to do an intense workout, when would you do it?**

    a.   Before 8:00 a.m. (1)

    b.   Between 8:00 a.m. and 4:00 p.m. (2)

    c.   After 4:00 p.m. (3)

8. **When are you most alert?**

    a.   One to two hours post wake-up. (1)

    b.   Two to four hours post wake-up. (2)

    c.   Four to six hours post wake-up. (3)

9. **If you could choose your own five-hour workday, which block of consecutive hours would you choose?**

    a.   4:00 a.m. to 9:00 a.m. (1)

    b.   9:00 a.m. to 2:00 p.m. (2)

    c.   4:00 p.m. to 9:00 p.m. (3)

10. **Do you consider yourself...**

    a.   Left-brained, aka a strategic and analytical thinker. (1)

    b.   A balanced thinker. (2)

    c.   Right-brained, aka a creative and insightful thinker. (3)

11. **Do you nap?**

    a.   Never. (1)

    b.   Sometimes on the weekend. (2)

    c.   If you took a nap, you'd be up all night. (3)

12. **If you had to do two hours of hard physical labor, like moving furniture or chopping wood, when would you choose to do it for maximum efficiency and safety (not just to get it over with)?**

    a.   8:00 a.m. to 10:00 a.m. (1)

    b.   11:00 a.m. to 1:00 p.m. (2)

    c.   6:00 p.m. to 8:00 p.m. (3)

13. **Regarding your overall health, which statement sounds like you?**

    a.   "I work out a lot, eat well, and avoid the bad stuff." (1)

    b.   "I try to do the right things. Sometimes, I succeed." (2)

    c.   "I hate exercise and love cheeseburgers, and that's not going to change." (3)

14. **What's your comfort level with taking risks?**

    a.   Low. (1)

    b.   Medium. (2)

    c.   High. (3)

15. **Which do you consider yourself?**

    a.   Future oriented with big plans and clear goals. (1)

    b.   Informed by the past, hopeful about the future, and aspiring to live in the moment. (2)

    c.   Present oriented. It's all about what feels good now. (3)

16. **How would you characterize yourself as a student?**

    a.   Stellar. (1)

    b.   Solid. (2)

    c.   Slacker. (3)

17. **When you first wake up in the morning, are you . . .**

    a.   Bright-eyed. (1)

    b.   Dazed but not confused. (2)

    c.   Groggy, eyelids made of cement. (3)

18. **How would you describe your appetite within a half an hour of waking?**

    a.   Very hungry. (1)

    b.   Hungry. (2)

    c.   Not at all hungry. (3)

19. **How often do you suffer from insomnia symptoms?**

    a.   Rarely, only when you're adjusting to a new time zone. (1)

    b.   Occasionally, when you're going through a rough time or are stressed-out. (2)

    c.   Chronically. It comes in waves. (3)

20. **How would you describe your overall life satisfaction?**

    a.   High. (0)

    b.   Good. (2)

    c.   Low. (4)

SCORING:

19 to 32: **Lion**
33 to 47: **Bear**
48 to 61: **Wolf**

My chronotype [Lion, Bear, Wolf, Dolphin] is _____.

Lion

Like their animal counterparts, human Lions are predawn hunters. They wake up hungry and burst with radiant energy that maintains itself throughout the morning and into the early afternoon. But around 5:00 p.m., their energy starts its steep and rapid decline. Ideally, they'd be in bed by

9:00 p.m., and they have to psych themselves up to go to parties and social events at night. They prioritize their health, exercise regularly and competitively, avoid drugs and alcohol (well, maybe just one glass . . .), and eat well. Among all the chronotypes, they have the lowest BMI.

Optimistic, ambitious, and emotionally stable, Lions are fearless, confident, and driven. However, their big-picture outlook makes it hard for them to notice subtle emotional cues in others. They love to fix things, broken or otherwise, which can cause tension in relationships.

Born leaders, Lions are introverts who might feel lonely at the top. They are goal-oriented agenda setters and list makers. They pounce on problems, find solutions, and relish playing the hero who saves the day. Cognitively, they have total clarity in the early morning, when most of the world is waking up. When their concentration lags in the afternoon, their creative energy kicks in.

As soon as Lions reach a goal, they look toward the next, and the next. The drive that propels them up the corporate ladder or CrossFit level board motivates them *and* depletes them. If Lions can adopt a restorative practice—stretching or meditation—they'll have even more energy for achievement.

Bear

Like their animal counterparts, human Bears are diurnal, meaning they are active in the daytime and restful at night. If they could, they would crawl into a cave and hibernate all winter. Waking up is a long process of hitting the snooze button and dragging themselves out of their warm beds. They wish they could get a few more hours (but rarely do). By midmorning, Bears are alert, but by midafternoon, their energy plummets, only to rise again in the early evening and then slowly decline until bed. They love food and would gladly snack all day (and night). Too much noshing, though, results in an above-average BMI. Bears are often weekend warrior exercisers and Sunday afternoon nappers, which sets them up for muscle soreness and injury, as well as Sunday night insomnia.

Bears are friendly extroverts. They're happiest when surrounded by people. When they spend too much time alone, they get antsy and increasingly anxious. While some do experience some social anxiety, their peak social energy conveniently coincides with happy hour. In relationships, even-keel Bears tend to avoid conflict and hope personal problems will work themselves out. Their moods reflect their life circumstances. When things are good, Bears feel good. When things are bad, Bears get anxious and depressed.

Team players, Bears think and work best in groups, but there are some who prefer a solo workspace. If meetings and brainstorming sessions take place before lunch, Bears shine. After lunch, they aren't as sharp, but their uptick in afternoon creativity and charm makes up for it.

Wolf

Wolves in nature come alive when the rest of the world goes to sleep. Human Wolves are at their most alert at sundown and don't feel tired until midnight or later. Their mornings usually pass by in a fog. Their bodies are up and moving, but their brains are still half-asleep. Most Wolves aren't hungry at breakfast and will drink copious amounts of coffee to clear that brain fog, though it doesn't work. But by the afternoon, they are ravenous and will make up for a missed meal or two by eating a huge dinner and snacking late into the night. Exercise? Wolves would rather drink wine, eat cheese, and debate philosophy until 2:00 a.m. They have the highest BMI of all chronotypes and are most prone to obesity-related conditions like diabetes and high cholesterol.

Wolves tend to be impulsive and creative. Their moodiness and pessimism (especially in the morning) can be challenging for their partners and families. But if there is an issue, they won't shy away from it. Wolves aren't afraid to talk things out until the problem is fixed. Pleasure seekers, Wolves are happiest when they are trying new things and indulging in everything life has to offer. Although they love to party, Wolves need a lot of alone time, too.

Highly creative, Wolves spark with bright ideas all day, but they are only able to concentrate effectively after 2:00 p.m. Since Wolves don't feel fully alive unless they are experiencing a high degree of intensity, they exhaust their energy quickly and need to incorporate a restorative practice into their lives. But they might need convincing. You can almost hear a Wolf saying, "Meditation? But it's just so *boring*."

Dolphin

In nature, Dolphins are unihemispheric sleepers. Half of the Dolphin brain is awake to prevent drowning and watch for predators while the other half sleeps. Human equivalents are insomniacs who can relate to feeling half-awake and half-asleep 24/7. A little neurotic and risk-averse, Dolphins tend not to be recreational drug or alcohol users, but they'll comply with their doctor's advice and take prescribed medication. They tend not to overindulge in food, and they have a limited number of sexual partners. Their BMI tends to be lower than average but not because they are obsessive exercisers. They burn calories from fidgeting and worrying.

Dolphins are unguarded in their relationships and tend to be caring, dedicated, and attentive parents and partners. But since they are so conflict-averse, they let small issues go until they become major problems, which can be energy draining and stressful in relationships.

Highly intelligent, Dolphins mind the details and will tinker with projects until they meet their impressive standards. They hit their logistic thinking peak in the evening and their creative peak in the midmorning. They are often too wound up at night to slow down and relax before bed, but they can make changes to their routine to quiet their active minds so they can get some decent sleep.

WHAT'S YOUR POWER PROFILE?

Your Power Profile is your metabolic speed (the most pertinent body-type factor for energy) of Slow, Medium, or Fast, plus your chronotype of Dolphin, Bear, Wolf, or Lion. For example, if you are a Slow metabolism endomorph and a Bear, your Power Profile is Slow Bear.

Just draw a line connecting your metabolic speed and your chronotype:

Slow   Bear
Medium   Wolf
Fast   Lion
          Dolphin

My Power Profile is _____.

Data nerds, prepare for a thrilling next few paragraphs. Everyone else, skim.

We were curious about percentage breakdowns of each Power Profile in the population, so we sent out several email blasts to Michael's online community and asked people to take the same quizzes you just did. We received more than 5,000 responses and compiled the data to find the most common Power Profile types.

According to Michael's previous research and the prevailing research in the field, the chronotype breakdown in the general population is 50 percent Bear, 20 percent Lion, 20 percent Wolf, and 10 percent Dolphin. Half the respondents in our new survey were Bears, which makes sense. We found more Dolphins in our data than in general population breakdowns, which also makes sense since they follow the Sleep Doctor blog and are more likely to have insomnia issues. Our breakdown: 50 percent Bear (2,630 completed surveys), 33 percent Dolphin (1,726), 10 percent Wolf (550), and 7 percent Lion (350).

Although the vast majority of people on the planet have either a Slow or a Medium metabolism, two-thirds of our respondents were Medium and nearly one-third Slow. Our Fast metabolism respondents accounted for around 3 percent, smaller than the general population of 5 percent. According to a recent study of nonathletes published in the *International Journal of Physiotherapy and Research,* 95 percent are either Slow or Medium types.[9] The lanky Fast type is what many of us aspire to be, and yet nearly *all of humanity can never transform into it.*

Why are there so many Slow types? We evolved to store fat readily so that during famines, we won't starve. This made sense in caveman times, when the food supply was by no means secure. One bad winter, and the tribe would die out if it weren't for that ability to store fat. But now we are (for the most part) constantly surrounded by food and burn far fewer calories than our prehistoric ancestors.

Eighty percent of our Lions had a Fast or Medium metabolism, which was predictable, since they tend to be conscientious about everything, especially their exercise habits. They tend to "eat to live" rather than "live to eat" (ahem, Bears and Wolves with their lusty appetites). Since Dolphins also tend to be wired type A neurotic fidgety types, we expected to find Fast types in this group, and did.

We did *not* find many Fast Bears and Wolves! Ninety-nine percent of those chronotypes in our survey were Slow or Medium. Since the percentage of Fast Bears and Fast Wolves is so small, we have not created Power Protocols for these types. We grouped the Power Protocols for Medium and Fast Dolphins together because their sleep and movement schedules are identical. We also grouped Medium and Fast Lions together for the same reason.

OUR POWER PROFILE DATA BREAKDOWN

	Medium (muscular)	Slow (curvy)	Fast (lanky)
Bears (50% of survey respondents)	67%	31%	<2%
Wolves (10%)	65%	33%	<2%
Lions (7%)	75%	20%	5%
Dolphins (33%)	70%	25%	5%

Living against your genetically predetermined Power Profile is the reason you're so tired all the time. But moving, eating, and sleeping in sync with your profile by sticking to a daily schedule—what we call Power Protocols—will give you renewed energy, like a brand-new pack of shiny batteries. You'll stop saying, "I'm so exhausted!" and start feeling the lift and power to move your personal mountains.

You might be tempted to skip ahead to Part II of this book, read the highly detailed chapter on your Power Protocol, and start the program immediately. But if you can bear (Lion, Wolf, and Dolphin) with us (hey, Michael is a dad, and he gets to make the occasional dad joke), the next several chapters cover a lot of ground on energy and what fuels us. We have the research on how sleep, movement, eating, and emotions affect your energy level, and we can show you how to squeeze more energy out of everything you do, maximize gains, and minimize drains.

We've got the goods not only for you but for your partner, parents, siblings, colleagues, bosses, friends, kids, and anyone else you care about. Everyone in your life should take the quizzes to discover their Power Profile. When you know their energy type, you can gain so much insight into who that person is and how they function. You can also check for potential compatibility issues you might have with them, like mismatched sleep and eating schedules and overall energy level. Conflicts definitely exist, but when you know what they are, you can prepare and adjust.

With all these resources, you'll be in control, supercharged up, and allowing your natural energy to take you wherever you want to go.

Everything you need to feel fully alive is quite literally in you already.

TAKEAWAYS: WHAT'S YOUR POWER PROFILE?

- The four chronotypes are Lion (early riser), Wolf (late riser), Bear (in between), and Dolphin (insomniac).
- When you follow your natural-born chronotype's schedule, all your systems run more effectively and energetically.
- The three main body types are curvy endomorphs, with a Slow metabolism; muscular mesomorphs, with a Medium metabolism; and lanky ectomorphs, with a Fast metabolism.
- By accepting your chronotype and body type, you can become the best and brightest version of who you already are and enjoy the physical and emotional energetic benefits that come from being in sync with yourself.

- Your Power Profile, a combination of your body type and your chronotype, paints a detailed portrait of your body, habits, and personality.
- Once you know your Power Profile, you can reorganize your life according to your Power Protocol, a detailed daily schedule, for maximum energy for *you*. It's your map to a speedier metabolism, better sleep, and a sexier, stronger, more confident you.

TO DO RIGHT NOW!

1. Take the body-type quiz and identify your metabolic speed: Fast, Medium, or Slow.
2. Take the chronotype quiz and identify your chronotype: Lion, Bear, Wolf, or Dolphin.
3. Figure out your Power Profile by combining the two — e.g., Fast Lion (like Stacey) or Medium Wolf (like Michael).

CHAPTER 3

Resting Energy

Sleep is the transfer of energy. That might seem odd because when you're out, you're just lying there. But a lack of movement doesn't mean that nothing is going on inside your body. In fact, sleep might be the most internally energetic time of your day.

Even the most powerful battery in the world is useless if it runs dry. The human body is a kind of rechargeable battery that energizes everything you do — cellular repair, digesting a meal, having a thought, or walking down the street. Your body battery keeps going and going, energizing you 24/7 for your entire life, no days off. And it's the only one you will ever get, so give it a rest!

Like your phone, your body battery needs to recharge every night. That's what sleep does. It's like plugging yourself into a power source so you can start the next day at 100 percent. If your body doesn't get enough time to recharge or the power source is weak, you won't get all the way to four bars, and then the next day, you'll function at a fraction of your capacity.

To energize your life, rest is *essential*. When people say, "I'll sleep when I'm dead," they might as well add "which will happen sooner than it needs to." Michael is the Sleep Doctor, and even he ran down his body battery without fully recharging…and paid the price for it. Rest is just. Not. Optional.

MORE OF THE GOOD STUFF

Upping your weekly energy score is possible only if you fully recharge at night with high-quality sleep for a sufficient number of hours. If you long to wake up like people in mattress commercials, throwing back the covers, beaming, dewy fresh, and ready to walk the dog, you need to have logged at least six-plus—ideally seven or eight—hours of undisturbed sleep.

Over the course of the night, you'll travel through different stages, starting with alpha and theta brain wave light sleep, then dipping slightly deeper as brain waves slow down, moving into deep, slow-wave delta sleep, and finally, into faster brain wave rapid eye movement (REM) sleep, when dreaming occurs.

• During deep, slow-wave sleep, the body restores and repairs, fights off infections and viruses, and produces an abundance of human growth hormone (HGH) to repair cells throughout the body—including skin, muscle, and bone cells. Sleep accelerates healing, even more than good nutrition.[1]

• Mental restoration happens during REM sleep. The brain's self-cleaning glymphatic system switches on and proceeds to sweep out neurotoxic waste products that build up during waking hours from your central nervous system by increasing the volume of cerebrospinal fluid in your brain and spinal cord.[2] Along with taking out the "trash," the glymphatic system aids in the distribution of glucose (sugars), lipids (fats), amino acids (proteins), and neurotransmitters (chemicals like hormones) that fuel the brain.[3]

• In the deepest stages of sleep, the brain restores important synapse connections and trims away unimportant ones. If you only get light sleep and not enough REM, you're missing out. It'd be like washing your hair but not rinsing.

↑ **Getting sufficient hours for sleep-based memory consolidation and mental processing is a profound energy gain.** Science shows us that if we have a peaceful night's rest, our waking minds will be sharper.[4] By sharper, we mean having a better memory, being a better problem solver, managing your emotions, and being more creative. You won't flounder when you face tough choices or have to think on the spot. You'll call those shots and move on without squandering an iota of energy.

↓ **So much energy is drained in frustration and indecision caused by sleep deprivation.** Think of the last time you felt really tired and had to make an important decision. Speaking for ourselves, we've had exhausting mini meltdowns over something as small as what to cook for dinner after a bad night's sleep. You certainly don't gain energy when you're upset and confused. If you're sleep-deprived or disrupted (waking frequently during the night), you miss out on all that physical and mental rejuvenation. Lack of sleep robs your body of muscle mass and strength and reduces bone density and the production of new, strong cells. It suppresses your immune system (aka *no bueno*).[5]

In the run-down-and-worn-out state of body and mind, you're not going to feel like doing energy-gain activities, like sticking to your movement and healthy-eating schedule . . . and then the cycle of exhaustion-stagnation will only continue. A major cause of sleep deprivation is living against chronotype. But by waking and going to bed according to your Power Profile's circadian rhythm, you will fall asleep faster and have fewer awakenings. It's the best way to optimize your sleep so that even if you get fewer hours than you'd like, the quality will improve, you'll feel more rested, and you'll get all of sleep's restorative benefits.

SEVEN AND A HALF HOURS OF RECHARGING BLISS

So how many hours does each chronotype need per night?

Throughout our adult lifetimes, in order to feel *up,* most of us need between seven and nine hours of sleep per night. Five and a half or fewer hours, night after night, is an ass-drag guarantee.

The goal of enough nightly hours for your chronotype might seem like the impossible dream (as it were). People have lives, Netflix queues, and Candy Crush levels to clear. It's not realistic for most of us to hit that ideal. Some shoot for six or seven hours at night with a midday nap to close in on a total of seven and a half hours within a twenty-four-hour period. (More on naps soon.) That's enough to get Lions and Bears to a four-bar battery recharge. Extreme Wolves who can't sleep until 1:00 a.m. have a tougher time getting to seven and a half without napping, but naps can make it harder for them to fall asleep when they want to at night. Yes, Wolves have it tough.

Dolphins can do just as well on six hours total because their sleep is compressed and intensified. They do more with less (our heroes), and we'd rather Dolphins feel emotionally satisfied with six hours than worry about not catching as many z's as Bears. If Dolphins get six hours, they're *fine*. If they get more than six, they're *awesome*.

Now, *which* seven to eight—or a Dolphin's six—hours are optimal for you?

WHEN TO TUCK YOURSELF IN . . . ZZZ

Go to bed when your body wants you to, which is about an hour or two after sleepy-hormone melatonin secretions start flowing.

Lions: Your body starts secreting melatonin in midevening and screams at you to get in bed at 9:00 p.m., when the party is just getting started for the rest of the world. If you force yourself to stay awake until midnight, you might please your pals, but when your internal alarm goes off at 5:00 a.m. and your body won't let you stay in bed for one more minute, you'll be exhausted the entire next day. We suggest working up to a 10:00 p.m. bedtime.

Bears: The societally conditioned bedtime of 11:00 p.m. to midnight is perfect for Bears. Most of us are forced to live on a Bear's schedule. It makes sense, since half the population is Bears, and majority rules. If you're a Bear, your pineal gland starts secreting the hormone melatonin around 10:00 p.m., signaling to the body that it's time to shut down. You'll be ready to sleep at 11:00ish p.m.

Wolves: If you're a Wolf and got into bed at 11:00 p.m., you'd wind up lying awake for hours until your internal clock signaled sleep readiness.

Wolves are usually wide-awake then because their melatonin release is delayed for an hour or two. Getting in bed too early and not sleeping can trigger anxiety, which makes it even harder to shut down. Hours go by without rest, then the alarm goes off at 7:00 a.m.; the Wolf is at energy level 1 and can barely peel his eyes open. We recommend an in-bed time for Wolves of 12:30 a.m.

Dolphins: You have to judge how you feel. A hard rule for insomniacs is "Don't get in bed unless you are sleepy." For "tired and wired" Dolphins, "wired" often dominates "tired," and they don't feel the slightest bit sleepy at bedtime. If they get in bed anyway, not falling asleep right away could set off an anxiety/insomnia cycle that will keep them up all night. Since Dolphins are shooting for six hours, we recommend you put off getting into bed until midnight.

TUCK-IN TIME

- Lions: 10:00 p.m.
- Bears: 11:00 p.m.
- Wolves: 12:30 a.m.
- Dolphins: 12:00 a.m.

WHEN TO RISE AND SHINE...

In the wee hours, your body stops secreting melatonin and starts releasing the hormones adrenaline and cortisol. That process triggers a rise in body temperature, blood pressure, and heart rate. Each chronotype's ideal wake time is one to two hours after the cortisol tap opens. According to a recent study of serum blood levels,[6] the average morning cortisol jolt happens between 5:00 a.m. and 6:00 a.m. (typical for Bears). One outlier participant's cortisol peak happened at 2:00 a.m. (an extreme Lion), and another at 10:00 a.m. (an extreme Wolf).

We recommend that Bears wake up at 7:00 a.m., Lions at 6:00 a.m., and Wolves at 8:00 a.m. Just because these times are ideal doesn't mean they're realistic. Not every Wolf has the luxury of staying in bed that late.

The Dolphin's ideal wake time is harder to gauge with hormonal, temperature, and heart-rate markers. For insomniacs, cortisol, body temperature, and blood pressure are elevated throughout the night and actually drop in the morning—the exact opposite of the other three chronotypes. According to their biology, they are not designed to have deep, restful, continuous sleep. But if they are able to pass out by 1:00 a.m., they should be good to go six hours later, so their recommended wake time is 7:00 a.m.

RISE-AND-SHINE TIME

- Lions: 6:00 a.m.
- Bears: 7:00 a.m.
- Wolves: 8:00 a.m.
- Dolphins: 7:00 a.m.

> **Energize Hack:** If you don't need an alarm to wake up at the right time for you, you are getting sufficient quality and quantity sleep.

If the day starts off well and you continue to keep the recommended move/eat/sleep schedule that's right for you, you'll be energized and motivated to stick with it. But remember, for many people, *change takes time.* Your body has to adjust. Once you start feeling the power, when your efforts are validated with increased energy and high vibes, you'll want to reinforce the changes with more movement and a stronger commitment to quality sleep and a consistent bedtime/wake-time schedule (at least that is what happened to Michael and most everyone else!).

Consistent wake time is *the most important factor* for a strong, health-promoting routine for all types—especially on the weekends.

Many of us try to make up for the sleep debt racked up during the workweek by sleeping in on the weekend. Does it work? In a new study, researchers at the University of Pennsylvania had their participants sleep five hours during the week and eight on the weekends for six weeks straight.[7] The scientists measured many different variables, including mood, focus, and attention. And, as you've probably already guessed, *everything* got worse as time went on.

The cure to being exhausted Monday to Friday is *not* to sleep in on Saturday and Sunday. It won't work for two reasons:

- **You'll never catch up on the weekends.** If you get six hours per night and need eight, your debt from Monday to Friday is ten hours. So on Saturday and Sunday, you'd have to sleep for *thirteen hours* each night. Goodbye, weekend.
- **Sleeping in on Saturday and Sunday morning causes Sunday-night insomnia and "social jet lag."** We're all familiar with the phenomenon: You lie awake on Sunday nights because you didn't get out of bed until noon for the last two days and are ruminating about the workweek ahead, and your body thinks it's in a new time zone (social jet lag).

> ✸ Energize Hack: Avoid this trap by limiting weekend sleeping in to just one extra hour on Saturday and Sunday mornings.

Sunday-night insomnia sets you up for a week's worth of exhaustion. You'll wind up operating at a constant energy deficit.

OVERSLEEPING DRAINS ENERGY: TOO MUCH OF A GOOD THING

Oversleeping is not the luxury of Sunday snoozing. It's a condition called hypersomnia, and it actually drains energy. If someone sleeps for nine, ten, eleven-plus hours per night, still has to open her eyes with a can opener in the morning, and remains groggy and foggy throughout the day, it's a flashing warning sign. Those symptoms could indicate a sleep disorder such as obstructive sleep apnea, restless leg syndrome, or narcolepsy. Ironically, oversleeping and insufficient sleep have a similar laundry list of associated health issues, such as heart disease, diabetes, obesity, cognitive impairment, and overall higher mortality rates. Hypersomnia might be a symptom of a substance abuse problem or neurological disorder like Alzheimer's or Parkinson's disease, or a side effect of medication.

It's a major red flag for depression, with 40 percent of oversleepers having a mood disorder. Depressed women are at high risk for the condition, reporting unshakable fatigue all day despite nine-plus hours of sleep at night.[8] The answer to the forever question, "Which came first—the depression or the oversleeping?" isn't entirely clear. However, we know that each problem exacerbates the other.

It's possible an oversleeper is making up for a huge sleep deficit. Our bodies tend to seek the sleep we need. However, anyone who sleeps long hours for weeks on end but still feels so damned exhausted might have an underlying condition. If you are sleeping excessively and regularly *and* continue to drag through the day, it's time to call your doctor and figure out what's going on.

SLEEP COMPATIBILITY

We all love the idea of drifting off peacefully next to someone you love. But it's just as likely that the person on the other side of the bed tosses and turns, snores, hogs the covers, farts, keeps the light on for hours, smothers with cuddles, or doesn't cuddle enough. Because of these disruptions, according to a 2012 survey by the Better Sleep Council, 26 percent of American couples prefer to sleep apart.[9] And that'd be fine, if sleeping in separate beds (or bedrooms) didn't take a toll on intimacy and bonding. If overnight togetherness matters in your relationship, we can help.

• **Snoring.** Thirty-seven million Americans make snorting, gurgling noises when sleeping, per the National Sleep Foundation. The snorer should sleep on his side, avoid alcohol an hour before bedtime, try nasal-opening strips like Breathe Right or internal nasal dilators like Mute, and get checked for sleep apnea. The nonsnoring partner can use earplugs (noise level rated at 32 or below), use a sound machine, or deflect noise by building a wall of pillows.

- **Room temperature.** One of you likes it warm; the other prefers it cold. A compromise would be to split the difference and choose a temperature in between. The too-hot person can sleep naked, with just one blanket. The too-cold person can wear pj's and pile on the comforters. Some fancy mattresses have side-by-side adjustable heating, too. There are also new technologies specifically designed to help with this problem. Go to TheSleepDoctor.com and use the code ENERGIZE for special offers.

- **Mattress firmness.** If there is a serious incompatibility with this, save up and buy a mattress with side-by-side adjustable firmness (there are many to choose from), or buy a softer topper for one side and not the other.

- **Reading/viewing.** We recommend that all devices be turned off an hour before sleep time. But if you choose to ignore that excellent advice and want to read or use your iPad later than your partner, adjust devices to the lowest possible brightness and use headphones and *please* use blue-light-blocking glasses. The nonreader/viewer can use an eye mask.

- **A stillness discrepancy.** Some of us wake up in exactly the same position we fell asleep in. Some toss and turn and kick the covers while unconscious. A light sleeper might be roused by the partner's jostling. But we can't control how much we move when we're asleep. You can make the bed with separate blankets so the active sleeper won't tug off their partner's. And you can get a bigger bed and/or a mattress that prevents "motion transfer" by localizing movement to just one side of the bed.

- **A bedtime discrepancy.** Wolves and Dolphins shouldn't even *look* at their beds until midnight or 1:00 a.m. Those partnered with Lions or Bears who are in bed by 10:00 to 11:00 p.m. won't be able to have those in-bed pre-sleep cuddles. However, such couples can reach a cuddle quota on the couch earlier in the evening. At the early nighter's sleep time, the late nighter can sit in the bedroom on a comfy chair in the dark and have "pillow talk" until their partner drifts off. Then they can quietly leave the bedroom and go about their business until it's time for them to slip into the sack, quiet as a mouse, so they don't wake their snoozing partner. In the morning, the early-rising partner can be just as quiet and respectful of their late-riser partner's rest when they get out of bed. The care you show for each other's sleep can be just as bonding and loving as getting in bed and falling asleep together.

- **A cuddle discrepancy.** One of you can't sleep unless you're held; the other needs space. A couple compromise is called for: Agree that cuddle preference is personal and has nothing to do with the love between you (it doesn't), and then establish a ten- or fifteen-minute cuddle session that satisfies one person's closeness need and grants the other breathing room without guilt.

ENERGIZE YOUR WAKING BRAIN (BYE-BYE, BRAIN FOG)

Although cortisol secretions flow for seven hours before they plummet in the early afternoon, they're not always enough to clear away morning fogginess, especially if you're a Wolf or Dolphin attempting to get up too early. Melatonin secretions taper off for most of us by 4:00 or 5:00 a.m. For Wolves, though, melatonin is still pumping at 6:00 a.m., which suppresses cortisol. Waking up is nearly impossible when your body chemistry demands that you "Stay in bed!" What's more, going against your chronorhythm can be so disruptive to your hormonal ebbs and flows, it can actually *prolong* melatonin production. You will be physically up, but your brain is still half-asleep. This condition is called *sleep inertia,* more commonly known as brain fog.

Here are ways to cut through the fog and energize your waking brain:

- **Get some sun.** Twenty minutes of *direct* sunlight sends a message to the pineal gland that it's daytime and it can stop secreting sleepy-hormone melatonin, a circadian reinforcer. Not only that, but morning exposure to sunlight sets up a good night's sleep ahead,[10] so it clears brain fog while also preventing it. The application of this brain hack couldn't be easier. As soon as you can after waking, go outside and open your eyes. Don't look directly at the sun, though, and don't wear shades.

- **Drink water.** We lose about a liter of water every night while sleeping. Snorers and those with sleep apnea lose even more (fluid is expelled with every breath). Most of us wake up needing a 500 milliliters tall drink of water. There is a dangerous feedback loop between dehydration and sleep: If you're dehydrated, you might not get sufficient sleep, and not getting sufficient sleep causes dehydration.[11] The antidiuretic hormone vasopressin controls your body's water balance while you're sleeping, and it's released later in your sleep cycle. Short sleep interferes with that process. Other research has found that 80 percent of study participants with chronic kidney disease complain about sleep problems.[12] The dehydration–kidney disease connection, like oversleep and depression, is a chicken-egg situation. We don't yet know which came first. Hydration is linked to better sleep, though, and therefore less brain fog.

- **Cold showers.** Brace yourself: A blast of cold water in the shower will make you feel wide-awake. Anyone who's jumped into a mountain lake or tried to swim off the coast of Maine in July already knows that when you immerse yourself in chilly water, your heart starts pumping and you are far, far from sleepy. Cold showers increase oxygen levels, boost cardiac function, improve circulation, lower blood pressure, and boost immunity. They increase the release of feel-good endorphins, noradrenaline, and dopamine by up to 500 percent.[13] Start with a normal-temperature shower. Reduce the temperature so that it is colder than comfortable, and stand under the water (head included) for thirty seconds. Repeat the hot-to-cold cycle two or three times.

- **Stretch.** Wake-up stretching lifts mood, improves brain function, and increases *energy*. According to the research, low-intensity morning movement improves cognitive performance even if you had a bad night's sleep;[14] moderate-intensity movement improves working memory, attention, and decision-making.[15] Your morning movement Stretch session (read all about it in the next chapter) can help you remember where you put your car keys! Morning movement has also been found to decrease nighttime awakenings,[16] which means higher-quality sleep, a fully charged body battery, and mental alertness.

SOCIAL JET LAG IN A MUG

Let's talk about caffeine. According to the National Coffee Association, we're drinking more coffee than ever:[17]

- 70 percent of Americans drink coffee every week.
- 62 percent drink coffee every day.
- On average, Americans drink just over three cups per day.
- 21 million Americans drink six or more cups regularly.
- Americans drink coffee throughout the day, not only at breakfast.

According to a study of 2,259 Hungarians on the intersection of chronotype and caffeine consumption,[18] researchers found that a love of soda, energy drinks, and coffee was associated with eveningness (Wolves), as was caffeine use disorder, or an inability to kick the coffee habit and persistent use of it even when it was damaging. Morning types (Lions), on the other hand, preferred tea and had higher levels of well-being.

The threat of caffeine use disorder (CUD) isn't benign for Wolves. It can have serious consequences. In a forty-nine-day sleep study by researchers at the University of Colorado, participants were given either a double espresso or placebo and exposed to three hours of bright or dim light at bedtime.[19] Saliva samples were taken to test for melatonin levels.

Not surprisingly, the dim-light–placebo group's circadian rhythms were unaffected. The dim-light–espresso group's melatonin release was delayed by 40 minutes. The bright-light–placebo group's delay was 85 minutes. The bright-light–espresso group's delay was 105 minutes!

So if you have a cup of joe at 9:00 p.m. and are surfing the internet at midnight, you won't feel sleepy until nearly 2:00 a.m. Not only did caffeine and blue light block melatonin release; they shifted the participants' circadian rhythms by an hour or more.

Look, we're not trying to take away your latte. If we tried to separate Wolves, Bears, and Dolphins from their coffee, we might lose an arm. But we do need to tell you that drinking three or more caffeinated beverages (coffee, tea, energy drinks, soda) per day *does* increase your risk for sleep disorders.[20]

And now for the good news: If your daily one or two cups are timed correctly, they won't impact your sleep or energy at all.

- **Do not drink any coffee when you first wake up.** Your cortisol and adrenaline levels are amping up. Adding caffeine to that potent morning hormonal cocktail will do nothing to make you feel more "alert," but it will make you jittery. Drink 500 milliliters of water instead to rehydrate and energize.

- **Do have coffee with your midday meal or snack.** By 2:00 p.m., Wolves have a cortisol dip, meaning you'll feel a bit drowsy. Counteract it with a cup of coffee to regain energy and cruise into your peak period of the day.

- **Do *not* drink coffee after 3:00 p.m.** Although you might feel the effects of caffeine after twenty-five minutes, it takes far longer for your metabolism to clear it out of your system. It takes one hour to metabolize the alcohol from one drink, but it can take up to eight hours to clear away caffeine. You will feel your afternoon cup until nighttime (or at least your brain will). And if you have another cup after dinner? You will be caffeinated way past your bedtime. Researchers at Wayne State University in Detroit tested the caffeine effect by giving people 400 milligrams of it right before bed, three hours prior, and six hours prior.[21] Compared to the control group, all three of the dose-time groups had significant disturbances to their sleep.

SLEEP-ENERGY DRAINS

What drains sleep energy besides going to bed at the wrong time for your chronotype? Two major culprits: blue light from devices and drinking alcohol too close to bedtime.

Blue light is just a shorter wavelength of light on the spectrum. Sunlight is partially blue, and exposure to it during daylight hours boosts our energy and focus. The light emitted from energy-efficient LED bulbs is blue, and using them is very good for the environment. Digital light from our devices is also on the short end of the spectrum—and that's how blue light gets a bad rap. We spend so much time in front of screens—tablets, phones, TVs—that we are absorbing too much blue light, more than our endocrine system was designed for. A vast body of research has confirmed that the stream of blue light into our eyes blocks melatonin. It not only turns off the sleepy hormone but also turns on the awake hormone cortisol,[22] shortens sleep duration, negatively affects sleep quality, and leaves you feeling exhausted the next day.[23] What's more, overexposure to blue light has been linked to an increased risk for breast and prostate cancer,[24] obesity,[25] and diabetes.[26]

There are countermeasures you can take, like blue-light-blocking glasses,* which successfully reset circadian rhythms[27] and help you get a decent night's sleep.[28] If you want to try them, get amber lenses, which filter out high-energy-visible (HEV) blue light better than clear lenses. But you don't have to buy anything. Just shut down all your devices an hour before bedtime to offset the onslaught. Along with limiting HEV exposure, going off-screen encourages mental downshifting and disengagement at the end of the day. Don't rely on "night shift" mode, which turns your phone's backlight orange. The data shows it doesn't really do anything.[29]

> ⚡ Energize Tip: If you're unplugged, you won't wind up reading articles that get you all riled up when you're supposed to be winding down. Remember, the calmer your mind is when you fall asleep, the less anxious your dreams will be. So do not look at social media at night; it'll give you nightmares.

Alcohol consumption before bed prevents the body battery from

---

* If you are in the market for blue-light blockers, Michael has a partnership with Luminere Smart Eyewear and endorses premium blue-light-blocking glasses, which you can find online at SleepDoctorGlasses.com.

fully recharging. It's like hitting the sack with one bar and waking up with two instead of four. Alcohol is dehydrating, for one thing. A dry body is an energy desert. And sleep quality when tipsy? It goes down the drain. As we've said, the most restorative phase is during deep, slow delta-brain-wave sleep. Recent research has found that drinking before bed does put you into delta-wave sleep but simultaneously increases faster alpha waves.[30] So your brain is both asleep and alert at the same time. Hardly restful. It's the official definition of "wired and tired."

"Welcome to my world!" says every Dolphin everywhere.

We recommend that all Power Profiles abstain from alcohol within three hours of bedtime. Have your wine, beer, or cocktail with dinner, by all means, and then put the bottle away. A nightcap might make you pass out, but you'll wake up feeling like you barely slept.

> ⚡ Energize Tip: You may think wine puts you to sleep, but it won't *keep* you asleep. So rethink the Cabernet before bed and try soothing, sleep-inducing banana tea (see page 107) or guava leaf tea instead.

MICHAEL SAYS...

I don't drink a lot myself because alcohol disrupts sleep and circadian functioning.[31] Sure, I like a glass of wine with a great meal or a few beers on a hot summer evening, but I like being in sync with my chronorhythm even more. Alcohol reduces melatonin production by up to 20 percent.[32] It elevates adenosine, the chemical that builds up the longer you've been awake and can make you think you're more tired than you are, which glitches the master clock in our brains.

At least one in five Americans uses alcohol as a sleep aid.[33] I get it. After a drink or two, it is easier to fall asleep. But passing out is not the same as falling asleep. The irony is, according to a wealth of research, including a recent Finnish study,[34] even one drink impairs sleep quality. How they defined consumption:

- **Low intake.** Less than two servings per day for men and less than one serving for women decreased restorative sleep quality by 9.3 percent.

- **Moderate intake.** Two servings per day for men and one per day for women decreased sleep quality by 24 percent.
- **High intake.** More than two servings per day for men and more than one serving per day for women decreased sleep quality by 39.2 percent.

The sedative effects of alcohol diminish over time,[35] so you'll need more to get that sleepy, boozy feeling.

Alcohol before bed is dehydrating (an energy thief), makes you snore more (which disrupts the sleep of the person next to you), increases the risk of sleepwalking, exacerbates sleep apnea, and can cause insomnia.

The time of day you consume alcohol matters, too. Happy hour cocktails in the early evening are effectively metabolized by Bears and won't necessarily impact sleep. But brunch mimosas? They'll hit hard and disrupt your circadian functioning all day. And drinking a few glasses in the late evening is the worst for sleep. It sends a wrecking ball through your sleep architecture (the steady progression through the stages of sleep). You'll wind up significantly shortchanged of REM,[36] the stage of sleep when the brain's damaging proteins (aka the trash) are removed, for the first two ninety-minute sleep cycles of the night. When you wake up the next day, still fatigued from low-quality rest, your brain feels like garbage for a reason.

NAPS!

Mmmm, sleep lines, drool puddles, matted hair. You've got to love naps, especially on a Saturday afternoon while watching golf (Michael's personal favorite). They are just delicious, and so good for you! Even a ten-minute nap can reset your brain to give you sharper focus and creativity, lower stress, lift mood, boost energy for physical performance, and recharge your body...but only if you do it right.

The pitfall of napping for some, as Dolphins and Wolves know all too well, is that an hour of shut-eye during the day causes brain fog that afternoon and insomnia that night. For Dolphins, naps are totally counterproductive. You need to build up as much sleep pressure as possible during the day. Releasing a drop of that pressure with a nap will make it even harder for you to drift off at night. No matter how sleepy you feel, Dolphins, *no naps for you, EVER.*

Wolves can nap, if they keep it supershort—no more than a ten- to twenty-minute wolfnap. Bears and Lions have a smorgasbord of napping options to choose from:

- **The Executive.** To make wise decisions and sharpen your concentration, take a twenty-five-minute power nap post lunch.
- **The Nap A Latte.** This one is good for long road trips or study sessions when you feel sleep pressure building but only have limited time to unplug. Find a quiet spot where you can nod off. Drink a 180–240 milliliters cup of black drip coffee, quickly. Set your alarm for twenty minutes hence. Lie down, snooze, relieve that sleep pressure. Your alarm will go off just as the caffeine kicks in, so you'll feel refreshed and stimulated at the same time. Warning: Don't use this technique late in the day if you are sensitive to caffeine.
- **The Siesta.** A great tradition in Italy, Spain, China, Costa Rica, and many other parts of the world is lying down after lunch for rest-and-digest mode. We'd both like to see a tradition of afternoon napping take root in America. This 2:00 p.m. rest coincides with the drop-off in cortisol and what's known as the postprandial dip in blood sugar after the afternoon meal. Kids call it the JAMS, as in "just ate, must sleep." Forty-five minutes to an hour is long enough for your blood sugar to return to normal.
- **The Jetsetter.** We're both frequent fliers and have had to go directly from the airport to the convention floor or conference room many times. An in-flight nap staves off fatigue and has saved Michael's lecture performance more times than he can count. When traveling between time zones, an hour or two of napping can ease that process and make up for lost nighttime sleep. The trick of this nap is to rest on schedule with your arrival destination. Resting on the same schedule as the departure city will only delay your transition to a new time zone. Even a twenty-minute nap in flight can help you make it through the first couple of jet-lagged days and still allow you to fall asleep on local

time at night. If you want to use napping to actually fight jet lag, go to Timeshifter.com to download the app, which will save your sleep while traveling.

- **The Disco Nap.** Not just a throwback from 1978, this classic nap strategy boosts energy if you're planning a late night at the club or the library. Before you head out for the evening, take a ninety-minute nap at around 8:00 p.m. Warning: This will destroy your overnight rest, so use it only if you intend to pull an all-nighter. No matter how late (or early) you stay up, wake at your usual time. It'll be a long, tired, bad day, but by getting up on schedule, you won't throw off your rhythm for a whole week.

SLEEP + MOVEMENT: SUPERPOWER NAPS

Sleep and exercise go hand in hand. One aids the other and vice versa, which is why movement is a huge part of the Power Protocols. A recent study made a new discovery: Napping after exercise improves memory. Researchers divided 115 healthy participants in their twenties into two groups: exercisers who cycled for forty minutes, and non-exercisers. Both groups were then asked to memorize a series of pictures. After that, half the exercisers took an hour-long nap and half were told to remain awake. Non-exercisers were also divided into two groups with the same instructions. Next, all the participants were tested on which pictures they remembered from the earlier test. Of the four groups, the exercise-plus-nappers' memory scores were the most accurate.[37] As for how this neat trick can be used in a real-world scenario, say you need to energize your memory before a standardized test or a presentation. Two hours beforehand, take a brisk walk for half an hour. Next, look over your notes, and proceed to nap for half an hour. Then march into your test or meeting with confidence energy to burn. Hell yeah!

Lions can get away with a ninety-minute nap and still sleep at night. Unless traveling or trying to stay up late, Bears should avoid napping for longer than twenty minutes. The best time to nap in order to recharge your battery is seven hours after wake time, per Sara Mednick, PhD, author of *Take a Nap!* With that formula in mind, the ideal nap time for each chronotype is:

- Lions: 1:00 p.m.
- Bears: 2:00 p.m.
- Wolves and Dolphins: *Napping is not recommended.* You need to build up sleep pressure as much as possible during the day; relieving it with a nap is counterproductive to falling asleep when you want to at night.

STACEY NEEDS HER NAP!

When I lead a class at SoulCycle, I have about 140 eyeballs locking into mine and tapping into my source energy as a human being. There's an invisible rope that connects me to each person and they're tugging on it both spiritually and metaphysically. It's an incredible feeling of connection. Often, it's a complete charge because my students actually give positive energy back to me. But if a few negative people are suffering or if there's some bizarre metaphysical or spiritual misbehavior going on, it's an energy drain. Some days, at the end of class, I just need to lie down. When I'm zapped, I go down for a ninety-minute nap. When I wake up, I have a nice healthy meal or snack, and then I'm charged and ready to go.

The key is to accept and appreciate what my body is telling me I need. If I ignored that "Go lie down!" voice, I wouldn't be good for anyone, especially myself.

REST AND RECOVERY

Resting energy is not *just* about unconsciousness. It's about resting, relaxing, taking time off to give your body and brain a chance to recover and rebuild from exercise and the stresses of daily life. During an intense workout, your muscles are depleted of glycogen (sugar). It takes between twenty-four and forty-eight hours for that muscle fuel to be replenished. Working out before recovering won't increase muscle mass or strength; the lack of glycogen forces the body to consume muscle to power your workout. Not giving yourself a chance to recover is a counterproductive waste of energy, sweat, and time. Failing to slow down during and after an illness or a period of emotional intensity or intellectual labor is just as pointless. Prep for rest... it's a thing.

Rest helps you heal and bounce back more quickly, physically and mentally. It's as true for serious injuries, like brain concussions,[38] as it is for minor stressors like the common cold. The law of diminishing returns applies to fitness overdoers. Ironically, they push themselves so hard— coming to class with the flu or using exercise to conquer a rough patch— because they believe they'll feel better emotionally. But they wind up physically depleted and find themselves struggling for weeks to get back to where they were. (They always make a comeback, but the exhausting struggle was preventable!)

A lesser-known benefit to taking sufficient rest between workouts, deadlines, breakups, illnesses, and injuries is self-compassion. If you're rested, you're nicer to yourself. The gentler you are to yourself, the lower your stress. The lower your stress, the better you feel and sleep and handle life's curveballs. The energetic wheel continues to turn. The question is, are you on top of it, or being run over by it?

> ✷ Energize Hack: The more rest you get, the more control you have over the internal audio engineer in your head, the one that tells you you're not good enough or doing enough.

Say you're one of those people who vow, "I'm going to run a marathon before I die!" and proceed to exhaust yourself in training and by being self-critical. Running a marathon is not an appropriate goal for most people, yet they beat themselves up for failing to achieve this dream.

An energizing practice for all Power Profiles is to fit your dreams to who you are.

MICHAEL SAYS...

I've been working on my own negative self-talk lately to reduce the anxiety that caused my series of cardiac events. The audio engineer in my head told me to ignore my exhaustion and just hop on the next plane, work my fifteenth weekend in a row, run another 5K, even though I was in pain. I was waging a war inside my own head.

Sometimes, you're supposed to listen to the voice that says, "Push yourself." But other times, you need to stop pushing and take a break. I've been looking closely at why people aren't resting, and it has a lot to do with ruminative thought. They can't turn off their brains and quiet their critical internal dialogue.

How do you get a rest from your own negative thoughts?

Unless you're distracting yourself with drugs, drinks, Netflix, workaholism, or Doritos, you have to expend some energy by addressing it. The Energize program for increasing energy is about accepting who you are—your natural-born chronotype and body type. If you can do that, you'll feel more confident and stop wasting energy trying to be something you're not. After living like this for a week or two, lo and behold, you'll start to feel better in your body *and* your thoughts.

Being yourself is one thing; enjoying yourself is another. Some might say, "Well, I'm being true to myself as a workaholic." It might be hard for an overachiever to appreciate the benefits of rest. But your body doesn't care that your brain takes pride in working sixteen hours a day. Work hard at finding balance.

Acknowledging that you are human, and that all humans (even Lions) need to rest, is just another facet of self-acceptance. We are giving you guidelines about how to get adequate rest. But only you can judge and acknowledge when you need to do so, and then do it without guilt or shame. Resting gives you the gift of energy.

SLEEP RX

It's important that the body is energized and functioning at the highest level. To make that happen, we might need a little help from supplements.

Of course, don't waste another precious resource—money—buying a supplement you don't need. The next time you see your doctor, ask for a blood test to check your levels of key vitamins and minerals.

Magnesium: Almost everyone is deficient in magnesium, and that has tremendous impacts on the sleep cycle. Magnesium calms the central nervous system; one study found that 500 milligrams can reduce depression and anxiety symptoms[39]—a good thing at bedtime. It helps regulate melatonin, cortisol, and GABA, a brain-calming neurotransmitter—all related to sleep onset and quality.[40] The RDA (recommended daily allowance) is 350 milligrams for adults. You can add it to your diet via leafy greens, seeds, and nuts, but that can be hard to do. Stacey does the diet thing and Michael does supplementation.

> ⚡Energize Tip: When under physical or emotional stress, you're extra vulnerable. Resting energy is essential for recovering from any kind of stress, pain, or depression. When balanced with exercise, rest is the key to life.

Vitamin D: Vitamin D deficiency negatively impacts the duration[41] and quality[42] of sleep. Since sunlight and darkness guide our circadian rhythms, and sunlight is our best source of vitamin D, it's not too surprising that this vitamin keeps our inner clocks running. Most Americans need more of it. The National Institutes of Health recommends 600 IU per day, but Michael takes 5,000 IU per day. Or you can spend twenty minutes outside in direct sunlight.

Vitamin B$_6$: Getting more vitamin B$_6$ might help you with lucid dreaming or even just remembering your dreams.[43] A deficiency might cause insomnia and depression, since the vitamin aids in the production of both melatonin and serotonin. Taking too much B$_6$ can be toxic, so up your level with natural sources like milk, eggs, cheese, fish, and whole grains.*

---

* Sleep Doctor PM Night Time Formula, available at Amazon, is an oral spray that contains the ingredients valerian root, gamma-aminobutyric acid, magnesium, magnolia officinalis bark, melatonin, and 5-HTP. For wake-ups during the night, try Sleep Doctor PM Middle of the Night Formula, with no next-day brain fog.

Sleeping Pills

Sleeping pills can be useful to break the cycle of insomnia or when travel-ing. Many people successfully use these medications effectively and with-out danger. If you and your doctor have decided that this is an option for you, we respect that decision, understand, and agree. There are some incredibly effective and safe sleep meds available.

However, many people prefer to avoid dependency on them. In 2019, the US Food and Drug Administration began requiring the medications eszopiclone (Lunesta), zaleplon (Sonata), and zolpidem (Ambien) to carry its strictest "black box warning," alerting users to their risks in large print right on the packaging, specifically for women. A low dosage is sug-gested due to problems "clearing" the drug from women's systems. The risks are rare, but they are serious: accidental overdose; preventable injuries like falls, burns, drowning, car accidents, hypothermia, and carbon dioxide poisoning; and preventable self-injury, including suicide attempts.

There have been widespread reports of users sleepwalking, sleep-shopping, sleep-eating, and sleep-texting. No one wants to wake up with a dozen emails from retailers alerting you to overnight purchases you can't remember making. Or being told by a roommate that you got in bed with them or that you picked a fight with someone. Or finding out that you sleep-drove or turned on the stove and sleep-cooked (one of Michael's patients made scrambled eggs in her sleep and left the stove on!). Researchers don't fully understand the mechanisms that make such dan-gerous behaviors happen, but in many cases, *alcohol was combined with sleeping pills, a really bad thing to do.* The theory is that these medications put people into an altered state of arousal while in partial sleep. The drugs wear off, people wake up, and they have no memory of what they did while "asleep." Consciousness isn't a single state. It can be mixed, like a hybrid car that can run on gas and electricity. When some parts of the brain are "out," others might be "up."

Apart from hazardous risks, these meds have scary side effects as well, including dizziness, headache, nausea, diarrhea, depression,[44] mem-ory and performance issues, and dependency.[45] A 2018 review of three dozen studies identified a higher risk of death among users.[46] It doesn't take a lot, either. Per another study, taking hypnotic drugs just eighteen times in a year increased the risk of death threefold; the risk of developing cancer is increased by 35 percent for users vs. nonusers.[47]

What about over-the-counter sleep aids? Well, they're just not that effective and have more risks than benefits. The 2017 American Academy of Sleep Medicine review recommended against using them as a treatment for insomnia.[48] These drugs are usually antihistamines (allergy relief) and analgesics (pain relief). Although they do block "awake" brain receptors, they last for twelve hours. Unless you want to have that woozy feeling for half the day, it's not worth taking them. What's more, sleep aids have been associated with increased risk of dementia. In a University of California, San Francisco study, researchers asked 3,068 senior citizens without symptoms of dementia how often they used sleep medications, and then followed them for fifteen years.[49] The participants who claimed to take sleep meds "often" or "almost always" were 43 percent more likely to develop dementia compared to the participants who didn't take them.

MICHAEL SAYS...

Honestly, I prefer to use sleeping pills as a last resort while working with the patient's medical doctor. The all-natural techniques for getting to sleep—like progressive muscle relaxation and cognitive behavioral therapy—are actually more effective than meds[50] and have none of the risks. Talk to your prescribing doctor about safe use and alternative therapies.

To taper off sleep aids:

Four percent of Americans are taking sleep drugs. If you are one of them and would like to stop, do not act rashly and flush them down the toilet! Going cold turkey would likely cause withdrawal symptoms such as anxiety, lightheadedness, headaches, rebound insomnia, fatigue, nausea, vomiting, weakness, aches, and sweating, among others.

It's essential that you work with your doctor on a plan to taper off sleep meds slowly and methodically. Most doctors will reduce the drug amount by 10 to 25 percent weekly. It won't be easy, and it'll take time. But if you follow through, you can largely be free of sleep-drug dependence. See your doctor and ask, "What's the safest way to wean myself off sleeping pills? How long will it take? What are the risk factors? Can we make a plan today?"

What About Weed?

A very popular question submitted to the Sleep Doctor blog is, "Does cannabis help you sleep?" The answer is a lot more complicated than just "Go forth and vape."

We know that cannabis helps people fall asleep more quickly, whether they have insomnia[51] or not.[52] The natural chemical compounds in cannabis that help sleep are:

Cannabidiol (CBD): This compound has zero psychotropic effect but does reduce anxiety, relieve pain, promote mental focus, and reduce daytime sleepiness. You can find hundreds of lotions, drops, gummies, and edibles with CBD online and in stores. But the data is limited, and its dosage stats are vague. A recent study found that as little as 25 milligrams and as much as 175 milligrams can be beneficial for sleep.[53] Start at the low dose, and scale up slowly.

Cannabinol (CBN): Never heard of it? It's a pain reliever and an appetite stimulator with anti-inflammatory properties. Like CBD, CBN does not get you high, although it has a sedative effect that can make it easier to fall asleep. It turns out that CBN is actually oxidized THC (i.e., old weed), so it makes sense that it could make you sleepy. It's available in oil form as a sleep aid.

Delta-9-tetrahydrocannabinol (THC): THC is the compound that gets you the high, euphoric feeling, but it's a mixed bag re sleep. On the one hand, it appears to improve breathing during sleep—a potential therapy for obstructive sleep apnea. On the other, it reduces time spent in REM sleep (mental restoration) and increases delta sleep (physical restoration). So your body benefits from extra cellular repair, but your brain's trash-removal time is diminished. If you are prone to nightmares and post-traumatic stress disorder, though, THC can help. (Consult a doctor before you take anything, of course.)

Terpenes: These molecules are aromatic and give weed its distinctive smell and taste. They also affect mood, energy, alertness, and sleepiness. One such terpene called *myrcene* has sedative properties. Another, *caryophyllene,* relieves pain, stress, and anxiety. *Limonene,* a citrus-scented terpene, elevates serotonin, which in turn reduces anxiety, acts as a natural antidepressant, diminishes OCD behaviors — and reduces insomnia. One more: *linalool,* a lavender-scented terpene, is a superstar, lowering anxiety and depression, boosting immunity, and increasing adenosine, a hormone that helps us fall asleep.[54]

Unfortunately, no one seems to have solved the sleep/weed dilemma. While marijuana contains all kinds of substances that may help us up our sleep quantity and quality, the research in this area is severely lacking. Those who live in states where cannabis is legal can consult an expert at a medicinal dispensary to answer questions and suggest particular strains that are right for your individual needs. For sleep, look for a strain with some of the terpenes mentioned in this discussion, with a higher concentration of CBD than THC as a general rule.

TAKEAWAYS: RESTING ENERGY

- Your body is a rechargeable battery, and sleep is the power source to plug into to get back up to 100 percent.
- Quality shut-eye takes you through stages of light and deep sleep, when your body heals, and REM, when memory consolidates and your brain's trash is taken out.
- A well-rested mind is smarter and energized to make wiser decisions.
- A well-rested body is stronger, with an energized immune system.
- Lions and Bears should shoot for at least seven and a half hours of sleep per night (or a combination of night sleep plus daytime napping).

Wolves should avoid napping and shoot for at least seven hours. Dolphins should shoot for seven, but six might be enough for them.

- Consistent wake time—especially on the weekends—is *the most important factor* for a strong, health-promoting routine for all types, not just insomniacs.

- If you sleep for nine-plus hours per night and are still tired, it could indicate an underlying health condition or depression. Call your doctor.

- Sleep inertia is walking around with an awake body and a half-asleep brain. Beat brain fog by getting sunlight, hydrating, taking cold showers, stretching, and doing mind-clearing breathwork.

- Prevent insomnia by unplugging from blue-light-emitting devices and avoiding alcohol for an hour or two before bed.

- Napping can help make up for insufficient sleep at night and give you an afternoon energy boost—just not for Dolphins or Wolves.

- Rest after a workout increases the fitness payoff. Not resting diminishes it.

- Give yourself a mental break, too. Quiet negative self-talk to give your emotions a rest.

- Certain supplements can improve sleep quality. Consult your doctor about supplementing with magnesium, vitamin D, vitamin B_6, and cannabis products (where it's legal).

TO DO RIGHT NOW!

1. Make a list of all the ways your sleep habits are keeping you from getting a decent night's rest.
2. Start thinking about changes you can make to improve your sleep—and your sleep compatibility.
3. Talk with your doctor about what types of vitamin and mineral deficiencies you could have. Get tested, and then stock up on supplements that you actually need.
4. Take some time to decompress after a workout to maximize the energy gain.

CHAPTER 4

Moving Energy

To stop dragging ass, you have to get off of it first, then *move*. We all know (and have heard) some version of "sitting is the new smoking," about the damage to our health of being too sedentary. And yet, we still sit around for most of the day. The problem is, we don't always know just how sedentary we are. When we're engaged on our devices, suddenly hours go by and we don't even realize it. Our minds are active, so it kind of feels like we're busy and getting shit done; meanwhile our bodies are becoming one with the chair. That disconnect is hurting our health and draining our energy. Have you ever looked up from your screen, checked the time, and said to yourself, "Wait, hours have gone by since I last stood up? That can't be right"? That Apple watch can only tell you so many times to stand up before you actually do!

A recent study by scientists at the Washington University School of Medicine analyzed data from 2001 to 2016 about TV and screen use; in addition, they surveyed 52,000 Americans of all ages about their "sitting behavior" from 2007 to 2016. All age-groups reported double-digit increases in screen time / sitting from a decade earlier.[1] Sixty-two percent of kids twelve and under parked themselves in front of a screen for two hours of TV, games, or videos per day, as did 59 percent of teens and 65 percent of adults.

You're probably thinking, *Two hours? I'm on my computer or phone all day long.* When you factor in work-related ass-in-chair time, Americans are sitting most of the day, for hours on (rear) end. Very few of you have integrated the standing desk, treadmill desk, or exercise-ball chair furniture into your offices. If you have, you're one step closer to fighting fatigue.

And if being sedentary had no impact on your health—much less your energy—no one would be raising the alarm about it. But sitting *does* affect health. In a review of eight studies and 12,000 participants, Canadian scientists found that the tendency toward "sedentary behavior" vs. "movement behavior" is literally killing us. Light-intensity physical activity—just moving around more often—lowers risk of all-cause mortality (aka death by any reason), obesity, and heart disease in adults. The researchers concluded that periods of physical activity throughout the day should be included in any and all of today's public health guidelines.[2]

In November 2020, the World Health Organization released its recommendations[3] for physical activity: You have to move throughout the day to prevent disease and death. Specifically, the WHO recommends that all adults aged eighteen to sixty-four get 150 to 300 minutes per week of any combination of moderate and vigorous aerobic physical activity, sit less, and replace sedentary time with light movement. If you follow our program, you will hit those benchmarks.

Your individual Power Protocol includes recommendations for five daily movement sessions that last for five minutes each (not counting seconds of rest between movements), aka your Daily 5×5. Once you get into your new groove of doing your Daily 5×5, you will be shocked at how much more energy you have.

Our customized plan uses the *whole body* and the *whole day*—and it's fun and doable over a *whole lifetime.* The Daily 5×5 provides specific energy gains and health benefits based on your chronotype and bodytype combo. By moving throughout the day, *every day,* you'll build the flexibility, stamina, strength, and balance we all need, regardless of age and gender, to feel alive and connected to our bodies.

At first, it might feel strange and seem hard to do the Daily 5×5. But we will help you integrate it into your life, and we promise you that over time, it will feel completely normal. The movements are simple and addicting, the sessions are spread out, and they're brief. Mentally, the stress of carving out an hour to work out is totally eliminated. No matter how busy you are, you can spare five minutes to move—and by doing it, you'll feel pumped and ready to do more.

STACEY SAYS...

I don't believe in time off from activity. There might be less-active days, but I'm always moving. And because I'm always moving, I have the energy to keep doing it. It's like being a human flywheel. Just short bursts are enough to keep up your momentum for energy. You'll naturally get into the habit of craving movement, trust me.

One concern I hear from clients is that the five daily five-minute movement sessions can't possibly be enough. "It won't do anything for me." Listen, the truth is, you don't need to go hard to gain strength and endurance if you move consistently, multiple times a day. I've been an athlete my entire life, and I've never been able to do more than ten or twenty push-ups. During the summer of 2020, I gave free online Instagram workouts on Tuesdays and Thursdays that were only ten minutes long, at 10:00 a.m., 1:00 p.m., and 4:00 p.m. What I found is that doing three mini-workouts twice a week made me stronger than I'd ever been before. Now I can do almost one hundred push-ups, in one session! My increased strength didn't come from intensive weight training. It was from those short sessions online. I was as shocked as anyone at how little it took to change my body. At fifty-three years old, I have to tell you, it felt *incredible*.

After I tell people that story, they say, "I don't want to bulk up, though." All the movements I recommend are plyometric, meaning they're short bursts of effort and use only your own body weight for resistance (unless your mini-me climbs on your back). Any basic plyometric exercise—like pull-ups, sit-ups, push-ups, and vertical jumps—will increase muscle mass, but only to the extent that looks natural and appropriate for your body. You won't bulk at all. (If you do, look at what you eat. That might be the cause.) Most likely, your body will appear more streamlined than ever before, and a more powerful version of you will finally emerge.

WHAT ABOUT WORKING OUT?

If you have a fitness routine already, only you can assess whether it's draining your overall energy and causing body stress or giving you a bona fide boost. Take the demands of your life into consideration. You will see diminishing energy returns from pushing yourself too hard. Good workouts should not zap your energy and put you on the couch all day.

There is evidence that morning fitness leads to weight loss, if that's what you're after. For a 2019 study, researchers at the University of North Carolina at Chapel Hill enlisted 100 overweight inactive adults to commit to exercising five times per week for ten months.[4] They had the choice of doing their workout at 7:00 a.m. or 7:00 p.m. The early exercisers lost significantly more weight than the evening exercisers, even though all participants burned the same 600 calories per session. Taking a closer look at the results, researchers discovered that the early exercisers ate less and moved around more throughout the day than the late exercisers. We bet they were snacking due to low energy!

But if you have performance goals, work out in the evening. According to a 2019 Israeli study, metabolic and performance efficiency happens in the evenings.[5] Researchers put nocturnal mice on treadmills at two different times—at the start and end of their waking hours. The rodents' overall performance for intensity and speed was 50 percent better during "mouse evening" compared to "mouse morning." What's more, during "mouse evening," researchers detected higher levels of metabolites, chemical end products of metabolism that showed the mice burned more sugars and stored fats for energy at that time. Effective metabolism means more energy, and therefore a greater capacity to exercise. Human testing results lined up with the mice's. *Better exercise efficiency happens in "human evening," not "human morning."*

If you do decide to keep up a fitness routine in addition to the Daily 5×5, the best time to exercise for each chronotype is:

- **Lions:** Go for it at 6:00 a.m., when your adrenaline is flowing. Or, if you can, wait until 5:00 p.m., when you'll run faster, train harder, and be less likely to sustain an injury.
- **Bears:** The in-betweeners who work out before lunch or dinner will burn more calories and prevent postmeal snacking.

✳ Energize Hack: Vigorous exercise within an hour of bedtime might disrupt your sleep,[6] but moderate exercise—you can talk easily while doing it—won't ruin a good night's rest.

- **Wolves:** Late-to-bed Wolves' peak performance hits at 6:00 p.m. They should do strength training or yoga in the evening before dinner.
- **Dolphins:** Light sleepers can use exercise to jolt themselves awake at 7:30 a.m. To lengthen and deepen your short, shallow sleep, you can try training at 5:00 p.m. or restorative yoga at 10:00 p.m.

MICHAEL SAYS...

Exercise and sleep go hand in hand. One benefits the other. So the best time for everyone to exercise is after a night of seven to eight hours of quality sleep.

How quality sleep benefits exercise: It speeds up recovery time between workouts, increases endurance and exertion, improves speed and accuracy, and lowers the risk of injury.

How exercise benefits sleep quality: It shortens the time it takes to fall asleep, extends sleep duration, and increases time in physically restorative deep sleep.

Having this knowledge put me on a certain exercise path. I used to go for a run three times a week, usually at my Wolf-approved pre-dinnertime slot. But I wasn't moving nearly enough, especially when I was on the road. I'd find myself trapped in an airplane seat or car for hours, and then I'd be stuck in a hotel room. Apart from my runs, I was sedentary.

I started to come around to the idea that movement every day, throughout the day, was more important for my health than three intense workouts. Stacey told me many times to evolve on this point, but oddly enough, it never really sank in. And then I heard Tony Horton, the fitness trainer and creator of the P90X program, give a lecture for my men's group METAL International. At the time, he was sixty, looked thirty, and was in truly awesome shape.

Someone in the group asked, "Tony, how many days a week do you exercise?"

He replied, "I get asked this question all the time. Let me ask you, if you dieted three days a week, how much weight would you lose?"

It just hit me like a ton of bricks. If you want to change, you have to live that change every day. As Stacey has always told me, "Don't die-t. Live it."

When I cut back on running and concentrated on specific movements throughout the day instead, I started to lose weight, significantly lowered my stress, and felt so energized that I basically gave up coffee entirely. I just didn't need it anymore.

If you want your system to work at its best, follow our Daily 5×5 to double your energy.

MAKE YOUR MOVES

Your total movement time per day is only twenty-five minutes. Hardly any time at all. If you don't believe you have that much time to give to yourself, check your cell phone usage data. How much time do you spend on Instagram or Facebook? A recent study recommended no more than thirty minutes of social media per day.[7] If you spend an hour on FB or the Gram, cutting back by half and moving instead is a tremendously energizing win-win. According to a wealth of research, limiting social media time decreases depression and loneliness (huge energy drains) and raises self-esteem (an energy gain).

As we've said, your twenty-five minutes of low-intensity movements aren't taken all at once. You'll do a little burst here, a little burst there.

To collect data for your energy assessment score, you've already programmed your phone alarm to go off at five key times during the day: when you wake up, midmorning, midafternoon, evening, and right before bed. Not coincidentally, these are the same times each day when you need to get moving. (Yes, we thought ahead.) If you can't drop everything and start jumping when your alarms go off, wait for a convenient opportunity within twenty minutes.

All energy types will use five different kinds of movement:

Stretch is for flexibility.
Shake is the antidote to sitting for too long.
Bounce gives you insta-energy.
Build grows muscles for faster calorie burn.
Balance is for coordination.

As for which type of movements to do when, it depends on your Power Profile. All types begin the day with Stretch and end with Balance. Otherwise, your Shake, Bounce, and Build schedules depend on your chronotype.

Lions wake and Stretch. Since they laser-focus early in the day, they'll need to Shake it out midmorning and Build in the afternoon. Their energy falls off a cliff in the evening, so they will Bounce into a second wind when they need it most and Balance during Power-Down Hour before falling into bed.

Bears will start the day with Stretch. They'll Shake it out midmorning, Bounce to reset in the afternoon, Build their muscles in the evening, and Balance during their unplugged Power-Down Hour, which starts one hour before bed. This is the basic progression for most of humanity.

Wolves will wake up and Stretch. Midmorning, they'll Bounce to clear away lingering brain fog. Their Shake-up happens in the afternoon when they fall into the concentration zone and forget to move for hours. They'll Build later in the evening when they feel most energetic. And finally, they'll Balance as part of their Power-Down Hour moments before bed to quiet their active minds.

Dolphins need to Stretch first thing after a night of fitful sleep and Bounce midmorning to jolt into wakefulness the half of the brain that's still sleeping. Their energy peak comes in the afternoon, which is a good time to Build. They'll Shake off an evening spike in nervous energy, and Balance during their Power-Down Hour to try to prepare for sleep.

As for body type, Slow types need a balance of cardio and strength, which the five types of movements will supply. In your Power Protocol, we've added an extra (optional) Bounce session for Slow types to get the metabolic fires burning.

Medium types might find that they enjoy one type of movement more than another, but we urge you to do them all. You might not be used to Stretching and Balance poses now, but you'll learn to love them.

Fast types will gain a lot from a whole-spectrum movement approach, too. In your Power Protocol, we've added an extra (optional) Build session, to make sure you maintain muscle mass.

Here's a basic schedule for the Daily 5×5:

	Wake-Up	Midmorning	Afternoon	Evening	Pre-Bed
Medium Bear	Stretch	Shake	Bounce	Build	Balance
Slow Bear	Stretch	Shake	Bounce	Build	Balance
Medium Wolf	Stretch	Bounce	Shake	Build	Balance
Slow Wolf	Stretch	Bounce	Shake	Build	Balance
Fast/Medium Lions	Stretch	Shake	Build	Bounce	Balance
Slow Lion	Stretch	Shake	Build	Bounce	Balance
Fast/Medium Dolphins	Stretch	Bounce	Build	Shake	Balance
Slow Dolphin	Stretch	Bounce	Build	Shake	Balance

THE ENERGIZE DAILY 5×5 MOVEMENT GUIDE

Along with the descriptions that follow, we've posted video demonstrations—featuring us!—on MyEnergyQuiz.com of each move and dozens of others as well. In the upcoming chapters for each individual Power Profile, we will go into much greater detail about which movements to do throughout the four weeks of the program.

DISCLAIMER: It's a good idea to consult your doctor before you begin any fitness program, including this one. If you have any physical impairment that would prevent you from doing these movements, don't do them! If you have any pain while doing them, stop!

Stretch

Warm up your spine for a flexible, energized day.

Why? To get a bendy back. Having a flexible spine increases your range of motion, prevents injury, reduces pain and tension, improves posture and balance, and strengthens core muscles. Every nerve ending in the body is connected to your spinal cord. Stretching your spine puts more space between your vertebrae so those nerves aren't compressed.

Where? In bed or on the floor beside the bed. If the floor is bare, use a yoga mat. Wear loose pj's or do these poses in underthings (or in the nude!).

How long? Five minutes. At first, we recommend getting to know each movement by concentrating on one or two at a time. Eventually, you'll do a one-minute-each flow.

Level? Start with the easy version and depending on your comfort with it, level up.

Key points: As you stretch, inhale deeply and exhale slowly, and be sure to have positive visions in your mind. Hold or repeat each pose until the time allotted has passed.

1. Child's Pose

Easy. Kneel on the floor or in bed with toes together, knees wide apart. Lower your upper body so that your torso fits between your knees. Extend your arms in front of you in a Y shape with your forehead on the bed/floor. Allow yourself to sink into the pose.

Medium. Same as the standard Child's Pose, but instead of your arms in front of you, sweep them to lie alongside your body, palms up, with your forehead resting on the floor or bed.

Hard. In standard Child's Pose, walk both of your hands toward the right to get a left-side stretch, gentle and smooth as you move. Hold for five breaths. Return to center. Then walk your hands toward the left for a right-side stretch; don't rush.

2. Cat-Cows

Easy. On all fours on the bed or floor, inhale, drop the belly, move your heart forward over your wrists, and lift your head up. This is "cow" (mooing optional). As you exhale, arch your back and drop your head into "cat" (meowing optional). Press the tops of your feet down. Inhale to return to cow. Exhale, arch into cat. Move with your breath through ten rounds, deepening each time.

Medium. Do ten rounds of standard Cat-Cows. Then come back into an all-fours neutral-spine Tabletop Pose. Lift your left arm toward the ceiling, and reach. This twists the spine. Hang for five breaths. Then put your left hand back on the floor and return to all fours with a neutral spine, and repeat on the other side.

Hard. After ten Cat-Cows and the spine twist on both sides, return to a neutral spine on all fours. Tuck your toes under, and then lift your butt up and back into Downward Dog Pose. Keep your knees bent slightly. Let your head hang and shake it "yes" and "no."

3. Sphinx

Easy. On your belly, put your palms facedown on the bed on either side of your face. Lift your torso and gently arch the back, eyes forward, looking regal like the Egyptian Sphinx. Hold for five breaths, and lower for one breath.

Medium. Come into a standard Sphinx Pose and activate your leg muscles by clamping them together. Hold for five breaths and lower for one breath.

Hard. Come into Sphinx with strong legs, and then, slowly, press into your hands and straighten your arms into Seal Pose for a deeper backbend. Go as high as is comfortable and hold for five breaths.

4. Dragonfly

Easy. On your back, extend your arms above your head and stretch your body by lengthening through the hands and feet. Hold for five slow breaths and release.

Medium. Roll onto your stomach and do the same full-body stretch. Gently lift your feet and hands off the bed. Hold for five slow breaths and release.

Hard. On your belly, lift your arms, legs, and chest off the bed as high as you can go so only your hips and belly are on the bed in Superman Pose. Hold for five slow breaths and release.

5. **Cannonball**

Easy. Lying on your back, bend your knees and bring them toward your chest. If possible, wrap your arms around your shins and squeeze. Hold for five breaths and release.

Medium. Bring your knees to your chest, and then let your legs fall to the right side while your shoulders remain flat on the bed. Bend your arms at the elbows with your hands in the air (like the arms of a cactus), turn your head to the left, and sink into the spine twist. Hold for five breaths. Then bring the knees back to center and switch sides.

Hard. Bring your knees to your chest, then grab the outside of each foot with your hands so your legs splay open, knees bent into Happy Baby Pose. By straightening one leg at a time, rock and roll from side to side to get a lower back massage.

Shake

Shake off the dust from sitting too long.

Why? To loosen stiff joints. After sitting for an hour, your body fluids stagnate. Lymph pools, blood flow slows down. Shaking it out sends blood and oxygen into your hips, shoulders, neck, and lower back. It's like applying WD-40 to the muscles and joints that get super tight from being in the same position for a long time. Along with boosting circulation, these movements undo tension and decrease stress.

Where? Deskside or tableside.

How long? Five minutes. At first, we recommend getting to know each movement by concentrating on one or two for the full five minutes. Eventually, you'll do a one-minute-each flow.

Level? Start with the easy version and depending on your comfort with it, level up.

Key points: Take deep, slow breaths. And keep your abs engaged! Repeat each movement until the time allotted has passed.

1. Neck Looseners

Easy. Stand up. With your feet shoulder width apart and one hand on a desk or table, look straight ahead, chin parallel to the floor, in a neutral head position. Move your gaze to the left as far as you can without discomfort. Then move through neutral to the right as far as you can go.

Medium. Standing up, with shoulders back, come to a neutral head position. Place two fingers on your chin, moving your jaw and head back. Count to five in Latin (just kidding). Then relax the head and lift the chin into a neutral position, keeping your shoulders down and back. The chin tuck lengthens your cervical spine and helps realign your posture.

Hard. Standing with a neutral spine, extend your right arm to the side palm up. Bring your arm over your head and rest your right hand on your left ear. *Gently* and slowly pull your head with your right hand so your neck bends to the right. Count to five in Japanese (just kidding). Return to a neutral position and drop your right hand. Switch sides.

2. Arm Circles

Easy. To unfreeze shoulder joints, stand up and extend both arms out to the sides in a T position, palms down. Make small circles — the diameter of a salad plate — with your arms. Do ten clockwise and ten counterclockwise circles.

Medium. Go bigger and make dinner-plate-size circles. Do twenty clockwise and twenty counterclockwise.

Hard. Go even bigger, with cheese-platter-size circles. Do thirty in both directions.

3. Leg Swings

Easy. Stand up, feet shoulder width apart, and place an anchoring hand on a desk or table. Put all your weight on your right leg and swing the left leg back and forth like a pendulum. Then switch sides.

Medium. Stand up and anchor yourself with your hand on the desk for lateral (side-to-side) leg swings. Extend your right leg in front of you with your foot six inches off the floor. With a firm quadriceps and tight core, swing your leg from side to side to flex the hip joint in a different direction. Switch sides.

Hard. Stand up and anchor yourself with your hand on the desk. Extend your right leg in front of you with your foot six inches off the floor. Make small circles with your leg, twenty-five clockwise and twenty-five counterclockwise. Switch sides.

4. Crescent Bends

Easy. Standing, reach for the sky with your arms over your head and interlace your fingers. Stretch toward the ceiling with your whole body for five slow breaths. Then tilt your arms to the right and send your hips to the left to make a crescent shape. Hold for five. Come back to center, and tilt in the other direction to make a crescent on the left. Hold for five. Return to center and drop your arms to your sides.

Medium. Reach for the sky again, and after making a crescent on each side, return to center. With a slight bend in your knees, fold forward over your legs and touch your toes. While breathing slowly, gently try to straighten your legs. Hold for five. Come back up to standing.

Hard. Reach for the sky plus Crescent Bends. With the arms overhead, do a gentle backbend by pushing the hips forward and reaching toward the wall behind you with your hands. Hold for five. Come back to standing straight and then dip into a forward fold. Come back to standing and lower your arms.

5. Trunk Twists

Easy. Standing with feet shoulder width apart, bend your elbows with your hands up (like the arms of a cactus). Then twist your upper body to the left and right, keeping the arms on the same plane as your shoulders.

Medium. Standing with your arms outstretched in front of you and with slightly bent elbows, twist to the left and punch the air with your right arm. Twist to the right and punch with your left arm.

Hard. This time, instead of keeping your legs straight, do a mini-squat with each twist. Bend your knees as you twist and punch, and straighten them as you come to center, then dip again as you twist and punch on the other side.

Bounce

Get your heart pumping.

Why? For an instant energy jolt. Jumping unleashes a blast of feel-good dopamine, serotonin, and energizing adrenaline. It's better than a shot of espresso or a candy bar for quick energy when you need it most. Bouncing increases heart rate and blood flow, floods your body with fresh oxygen, burns calories like a wildfire, improves coordination, strengthens bones,[8] and lowers the risk of injury. Per a recent study, jumping for ten minutes daily gave participants the same heart benefits as jogging for three times as long.[9] For Slow and Medium types who hope to lose weight, Bounce stokes your calorie-burning power by keeping your "excess post-exercise oxygen consumption" (EPOC) going, meaning you will continue to blast through calories for a while after you stop jumping. It's like getting twice the bang for your buck.

Where? Outside to get a vitamin D infusion. Otherwise, any space that's ten feet by ten feet.

How long? Five minutes. We urge you to get to know each movement *very well* at first, devoting all five minutes to just one or two. As you get comfortable, choose two for two minutes each with a one-minute rest in between.

Level? Start with the easy version and depending on your comfort with it, level up.

Key point: Repeat each movement until the time allotted has passed. *Keep your abs engaged!*

1. Jumping Jacks

Easy. Stand with your feet together, arms at your sides. Jump so your feet are shoulder width apart while lifting your arms over your head into starfish position. Then jump to bring your legs together and your arms back down to your sides. If your boobs are flying, cross your arms over your chest.

Medium. Do a standard Jumping Jack. When you land in the straight-line position, add a quick Trunk Twist to the left and right.

Hard. Do a standard Jumping Jack. Jump out, squat. Jump back in.

2. Burpees

Easy. The standard burpee is pretty challenging itself! As you get used to the move, it'll get easier. Start by standing upright. Reach down to put your palms on the floor. Hop back into a plank, and then hop forward. Stand up and do a vertical jump.

Medium. Do the standing, lower, and hop-back-into-a-plank sequence, and then add a push-up. Return to standing and do a vertical hop. The push-up can be with straight legs or with your knees on the floor.

Hard. A Gorilla Burpee is a push-up burpee plus a squat. The sequence is: stand up, lower palms to the floor, kick back into a plank, do a push-up, rise into a squat, then hop, then stand. Whew! It's tiring just to contemplate. If your blood isn't flowing after this, you might be dead.

3. Ice Skater Jumps

Easy. Stand upright and bend forward at the waist. Put your arms behind your back, hands resting on your sacrum. Lifting your right leg, jump laterally to the right and land on your right foot. Then jump laterally to the left, landing on your left foot.

Medium. As you jump right, swing your right arm back and your left arm forward; as you jump left, swing your left arm back and right arm forward. As you land on each side, hold the balance for a tick or two to get that core-stabilizing bonus.

Hard. Do a standard Ice Skater Jump with swinging arms, and add a toe touch. As you land on your right foot, reach down with your left hand and touch your right toes. As you land on your left foot, reach down to touch your left toes with your right hand. Bounce and Balance all in one shot!

4. Jump for Joy

Easy. Standing upright, bend your knees and launch yourself into the air. If possible, take a video of yourself to see how high you can get off the ground. Two inches? Three? Try to get higher each time.

Medium. Standing at the bottom of the stairs, jump onto the first step. Step down and jump again.

Hard. Jump onto the second step of the stairs, or onto a very stable bench or box. Step down carefully!

5. Skip!

Easy. Channel your inner eight-year-old and skip with *attitude* around the room.

Medium. Skip faster in place like your inner kid just mainlined cotton candy at a fair!

Hard. Power skip. This time, see how high you can jump with each skip, raising your arms for momentum and getting those knees up. Think amplitude.

Build

Strengthen your muscles to feel your power.

Why? To build and maintain muscle. We all need muscle to go anywhere and do anything, and unless we maintain what we've got, we'll lose it over time. Starting at age thirty, muscle mass dwindles 3 to 8 per-cent every decade.[10] The less we have, the greater our risk of injury, dis-ability, obesity, insulin resistance, osteoporosis, heart disease, joint stiffness, and even decreased height.[11] For Slow body types, Build is the key to streamlining your body composition, if that's what you're after. For a recent study, researchers divided participants into three groups: dieters who didn't do any exercise, those who did only cardio four times per week, and those who did only strength training the same amount of time. Over eighteen months, the strength trainers lost 18 pounds, more weight than either of the other two groups.[12] Just like Bounce, Build keeps your body burning calories even after you've stopped doing the moves. The beauty of Build is that you will grow and maintain muscle when you'd otherwise be melting into the couch and snacking.

Where? Anywhere you have a couch or chair.

How long? Five minutes. At first, we recommend getting to know each movement by concentrating on one or two at a time. Eventually, you'll do a one-minute-each progression.

Level? Start with the easy version and depending on your comfort with it, level up.

Key point: Repeat each movement until the time allotted has passed. *About those abs? KEEP THEM ENGAGED!*

1. Squats

Easy. Place one hand on the back of the couch or chair as if it were a ballet barre. With feet hip width apart, squat halfway down to the floor, then come back up to the starting position.

Medium. Stand in front of the couch or chair so the backs of your calves are about four inches away. Bend your knees until your butt just touches the couch cushions, and then pop back up to standing.

Hard. You have to get off the couch to pee at some point. After you've finished your business, do a hover over the toilet seat and hold. The bath-room is your personal gym. Use any moment to Build and energize!

2. Crunches

Easy. Lie flat on your back with your knees bent and feet on the floor. Interlace your hands behind your head. Find a spot on the ceiling aligned with your chest and focus on it softly. Inhale deeply, then using your core, lift your head, neck, and shoulders forward and up while exhaling, and then lower to the floor while slowly inhaling. Listening to a podcast has never been more productive.

Medium. Lie on your back with your hands interlaced behind your head and your legs elevated so your shins are parallel with the floor. Crunch up and touch your left elbow to your right knee, then twist to touch your right elbow to your left knee. Lower back down. Try to hold the crunch for a two count on each side to force a slower, more concentrated movement. Lower down.

Hard. Lie on your back with arms and legs stretched out so your body forms an X. Bring your right hand toward your left foot, then your left hand toward your right foot, lifting your head, neck, and shoulders off the floor.

3. Dips

Easy. Sit on the edge of the couch or chair, feet flat. Place your palms flat on the edge of the couch or chair next to your hips, fingers facing out. Now move your butt forward off the couch so you are supporting your body weight with your hands and feet. Slowly lower your butt until it's six inches from the floor. Raise yourself, by pressing into your palms and using your triceps muscles.

Medium. Stand a few feet away from the front of the couch or chair. Put your hands on the edge of the seat. Keeping your body rigid, lower and raise yourself using only your arms to do a standing push-up.

Hard. Do a standing push-up as above, but this time, as you go down, extend your right leg backward into the air behind you so you're only supported by your left foot. Do a set and then switch sides.

4. Kicks

Easy. Lying on your left side on the couch or floor, point your right toes and lift your right leg up and down fifty times. Then flip over and do the left leg. Switch sides again, and this time, do leg lifts with your foot flexed toward your nose.

Medium. Lie on your back. Lift both legs about six inches off the floor using your core muscles. Then flutter-kick both legs, like you're doing the backstroke.

Hard. On all fours on the floor, extend your right leg with your heel pointed toward the ceiling and donkey kick behind you as if you were about to crush someone in the nuts! Make sure to lead with your heel and jam the bottom of that foot up into the air toward the ceiling with your core muscles tight.

5. Wall Sits

Easy. Go to the nearest wall. Slide your back down against it, with your legs making an upside-down L, thighs parallel to the floor. Hold for ten seconds. Return to standing and repeat. Your thighs will shake! Fight for it as long as you can.

Medium. Same as a standard Wall Sit, but this time, instead of just holding the sit, lift one leg up after the other, like you're marching.

Hard. Same as a standard Wall Sit, but instead of marching, alternate kicking each leg out in front of you. Keep your abs engaged.

Balance

Center your body and quiet your mind.

Why? To improve your coordination, which you'll need as you get older. Also, *to calm down.* You spend most of the day energizing your body and mind. At bedtime, leave the concerns of life on the figurative shelf so you can recharge your battery completely for the next day. During the program, all Power Profiles will begin the practice of a Power-Down Hour before bed to turn off screens, take a bath, read, talk about light topics, and disengage mentally to shift your body and mind from awake to sleepy. During this hour, Balance is better than any sleeping pill. When you are standing on one foot, concentrating on not falling down, you'll head off a rumination downward spiral that can delay sleep and cause anxious dreams.

Combine Balance with a breathing practice to multiply the calming effects. The lungs need to move, too! Deep belly breaths—inhale into the belly button for a count of four, hold for four, and exhale for four—stimulate the vagus nerve, which runs from your neck all the way down into your intestines. When stimulated, the vagus nerve switches off your fight-or-flight sympathetic nervous system and switches on your rest-and-digest parasympathetic nervous system. This biohack shuts down the flow of the stress hormone cortisol, which, in turn, increases the release of calming serotonin to augment the sleep-inducing effects of melatonin.

Where? Anywhere quiet where you can be alone. The bathroom, the bedroom, a walk-in closet.

How long? Five minutes. At first, we recommend getting to know each movement by concentrating on one or two at a time. Eventually, you might choose to do just one for the full five minutes, or to break up the session with more than one.

Level? Start with the easy version and depending on your comfort with it, level up.

Key points: Can you guess? It's about abs. And how they should put a ring on them and get engaged. Breathe slowly and deeply! Hold each pose until the time allotted has passed.

Note: *If you have any balance problems, anchor yourself with a hand on the wall or on a piece of furniture.*

1. Tree Pose

Easy. While standing, shift your weight to the left foot. Place your right foot on your left ankle or on your inner thigh above the knee. Put your hands in prayer at the center of your chest. Stare at a fixed spot on the wall in front of you. If you lose your balance, just try again. Switch legs.

Medium. Same as a standard Tree Pose, but with arms lifted overhead in a Y shape, like the branches of the tree. Hold and breathe. Repeat on the other leg.

Hard. After doing Tree Pose while standing on your right leg, lift your left leg out in front of you and raise it as high as you can. Try to get it parallel to the floor. Hold for as long as possible. Switch legs.

2. Figure Four

Easy. While standing, lift your right leg and put your right ankle just above your left knee to make a Figure Four shape. Place your hands in prayer at the center of your chest and look at a fixed spot in front of you. Hold while breathing into the belly, then switch sides.

Medium. Do a standard Figure Four, and this time, bend your standing leg slightly and hinge forward so your upper body is at a forty-five-degree angle to the floor. You will feel a gentle stretch in your hip. Keep your eyes forward; don't look down! Hold while breathing into the belly, then switch sides.

Hard. A yoga Toe Stand is a circus-trick pose, so don't feel bad if you can't do it. Make your Figure Four. Bend your standing knee and hinge forward until you can place your hands on the floor. Bend your standing leg further until you're balancing on the ball of your foot. Now, slowly lift your upper body and put your hands in prayer. It's *hard*. This is a good challenge. Try to push through.

3. Tippy Toes

Easy. Stand with your eyes fixed on a point in front of you. With your hands in prayer at the center of your chest, come up on your tippy toes. This is called Palm Pose. Hold and breathe.

Medium. Do Palm Pose, and this time raise your arms into a Y shape overhead. Hold and breathe.

Hard. Even a Balance move as simple as Palm Pose can become super challenging…with your eyes closed. Try it again but this time, once you're steady with your arms in prayer position, close your eyes. You *will* lose your balance. That's okay! Just try again.

4. Eagle

Easy. Stand and bend your knees. Lift your left leg and cross your left thigh over your right thigh. If possible, try to wrap your left foot behind your right calf. If you're thinking, *No way possible is that going to happen,* just hold with the left leg crossed over the right. Balance for as long as you can and switch sides.

Medium. Add the arms. After balancing with the left leg over the right leg (with or without the foot wrap), swing your left arm under your right and fit the right elbow into the crook of the left. Try to clasp hands and raise your wrapped arms a bit higher so your hands are parallel to your face. Hold until you fall out, and switch sides.

Hard. Get into Eagle Pose with crossed legs and wrapped arms, bend your standing knee deeper, and hinge forward from the hips while tightening your core, eyes forward. Hold and then switch sides.

5. Dancer

Easy. While standing, lift your left foot and bend the knee so your left heel is near your left butt cheek. Catch your left foot behind you with your left hand. Just balance on your right foot like this until you fall out. Switch sides.

Medium. From easy Dancer with a standing right leg and bent left one, raise your right arm to the ceiling. Slowly hinge forward at the hips, lowering your outstretched right arm as you press your left foot into your left hand. The harder you press backward, the better you'll keep your balance as you hinge forward. Hold and switch sides.

Hard. From medium Dancer, keep hinging forward until your body is at a forty-five-degree angle to the floor. Keep pressing your foot into your hand, and slowly raise it higher, until that foot is level with your head. This is extremely hard and takes strength, flexibility, balance — and a lot of practice. But once you find this balance, you will feel quite proud of yourself. It's worth the commitment. Remember, it's called yoga practice, not yoga perfect. Try, try again.

DAILY 5 × 5 RX

The little bursts to stretch the spine, unlock stiff joints, increase heart rate, build muscle, and center the body and mind are all you need to stay energized all day and sleep well all night. Some finer points about how to incorporate them into your life:

- **You have all you need.** No special equipment required. If you happen to have a foam roller, use it with Stretch. A mini-trampoline (they are inexpensive and available online) will reduce Bounce's impact on your joints. If you have a Pilates band, you can use it with Shake to intensify those moves.
- **You decide your level.** You have easy, medium, and hard levels for each move. Anyone at any fitness level can do all three. But on certain days, you might be in the mood for an easy session, and on others, you'll want to go hard. As you are getting to know each movement, start on easy and see how your body feels that day. Increase to medium and hard as desired.
- **You won't break a sweat.** You don't have to worry about needing to shower or change clothes in the middle of the workday. These movements are short, strategic energy bursts that deliver quick oxygen to inactive muscles and speed up your heart rate just enough to energize the body and clear the mind. You will not pour sweat. If anything, you'll get a rosy glow and/or a dewy shine. Think of it as natural makeup.
- **Hydrate!** After each movement session, drink a full glass of water (except after Balance; you won't want a full bladder so close to bedtime). Hydrate or die-drate! If you want energy to flow, flood your body with water. Broad spectrum CBD water by Akeso is a great option, too. It's hemp-powered recovery water that helps reduce inflammation while also providing relief for stress and anxiety. Use the code ENERGIZE at AkesoWater.com.
- **Do as much as you can and feel happy about it.** If you can't do five minutes each time, do two. If you can't do five sessions that day, do

two. Every minute of movement offsets a minute of sitting and will do you some good. And there's always a new session, and a new opportunity for an energy blast, coming up in a few hours.

TAKEAWAYS: MOVING ENERGY

- All evidence has found that we're using screens more than ever and moving less than ever.
- Being sedentary has a huge negative impact on our health and makes us more likely to die prematurely.
- New recommendations from the World Health Organization urge people to move more and to do both intense physical activity and moderate activity.
- More movement means better sleep, which means more energy, which leads to more movement . . . and so on to infinity.
- Doing the Daily 5×5 will boost energy and speed metabolism.
- For whole-body, all-day energy, you need to Stretch for flexibility, Shake stiff muscles and joints, Bounce to increase heart rate, Build muscles, and Balance for coordination.
- Each Power Profile has a specific Daily 5×5 plan based on your energy needs throughout the day.
- Once you get in the habit of moving, you'll never want to stop.

TO DO RIGHT NOW!

Familiarize yourself with the movements. You don't have to do them yet. Just read the descriptions and look at the videos. Okay, do a few if you want. They are kind of fun. We hadn't skipped in years!

CHAPTER 5

Eating Energy

Energy is exhaustible, as we all know when we can barely keep our eyes open. It needs to be replenished constantly. Our bodies are like cars, cell phones, or any technology that "runs" on fuel like batteries, gasoline, electricity, the sun for solar panels, and the wind for turbines. We need to keep injecting fuel into our supply line or we will be forced to coast through life until our tanks run dry, when we make a dead stop. The most important points to remember:

Sleep allows us to replenish our body battery.
Movement makes our battery stronger.
Fuel—what we eat—sustains our body battery.

The quality of our fuel, just like the grade of gas we put in our cars, determines how well our systems function. High-quality fuel keeps the engine clean and prevents breakdowns. We asked certified functional nutrition coach Sarah Wragge (co-host with Stacey of *The Way* podcast) for some broad guidelines that would apply to every Power Profile. Here's what she suggested:

- **Hydrate.** Most people don't realize how dehydrated they are, especially if they are avid exercisers. When we don't drink enough water, we can mistake thirst for hunger and overeat instead of giving our body what it really needs. We lose up to a liter of water just breathing while asleep, so start your day with 500 milliliters of water to rehydrate. Drink another 250 milliliters after each daily movement session (again, not after

Balance), and sip throughout the day. But don't drink during meals—it dilutes digestive enzymes. Have your last sip an hour before bed to avoid having to get up and use the bathroom. Over the whole day, shoot for 2 liters of water. Herbal tea counts, but caffeinated beverages do not.

- **Restrict acidic foods.** Your body has to maintain a neutral pH balance between acidic and alkaline. Foods that tilt it toward the acidic end of the spectrum are alcohol, coffee, dairy, meat, grains, sugar, fruit, honey, syrup, vinegar, and soy sauce. Then your body has to work hard to get back to neutral. It's wasted energy. What's more, an acidic body makes you crave more acidic foods. Sugar begets sugar; alcohol begets alcohol. But if you eat foods that tilt the body's pH toward alkaline—like leafy greens, cruciferous veggies, avocados, coconut oil, lentils, and beans—your body maintains balance more easily, and the sugar cravings stop. It's the ultimate food freedom.

- **Avoid added sugar.** Most people don't realize how much sugar is in foods like instant oatmeal, power bars, and low-fat yogurt. Sugar is added to everything. Even if you check food labels, food companies use code names that you might not realize mean sugar, like organic cane, maltose, dextrose, and brown rice syrup. The easiest way to avoid added sugar is to steer clear of processed foods entirely. You can't add sugar to spinach or an apple.

- **Consume more fiber.** Dietary fiber slows digestion and insulin response and makes you feel full. High-fiber foods are vegetables, hard fruits like apples and pears, and nuts and seeds like flax and chia.

- **Be choosy about fruit.** If you are trying to lose weight or have concerns about diabetes, avoid high-glycemic fruits like bananas, stone fruits, citrus, melon, and grapes, which raise insulin and blood sugar. Low-glycemic fruits like berries are preferable. Always eat fruit on an empty stomach, and never combine it with starch or protein or it'll ferment in your gut and cause gas and bloating. To slow the release of insulin, pair fruit with a fat. An apple with organic almond butter is a winning combination.

- **Limit animal protein.** You don't need animal protein with *every* meal. Sarah eats 80 percent plants and 20 percent lean animal protein. Be careful where your meat is coming from; if possible, choose local, organic, grass-fed, grass-finished, wild, and pasture-organic options. You eat what the animals eat, so their food sources are just as important as yours.

- **Limit grains.** Grains are acidic, inflammatory, and bad for your immune system. Complex carbs like quinoa, millet, and buckwheat are slow burning and fiber rich. They're seeds, not grains. Choose quinoa over pasta. If you must have bread, go with sourdough, which is easier to digest, or Ezekiel sprouted-grain bread. Limit all grains to 200 grams per day.

- **Say no to dairy.** Sorry. If you must, make sure your dairy is whole-fat, grass-fed, and organic. Choose goat or sheep cheese (like chèvre, feta, and Manchego) over cow products. For milk, try alternatives like oat, almond, or coconut milk. For butter, substitute with ghee (clarified butter).

- **Say yes to veggies.** The sky's the limit with green leafy vegetables, cruciferous veggies, and pretty much anything at the farmers' market. Limit potatoes and starchy squash to 140 grams per day.

- **Go nuts.** Choose nuts in their raw form, ideally local and organic. Avoid all roasted, salted, sugar-coated nuts. Cashews and peanuts tend to have more mold and can cause inflammation, so avoid them, too.

MICHAEL SAYS...

Just wanted to add some info about sugar and sleep. According to a recent study published in *The Journal of Clinical Sleep Medicine,* high sugar intake is linked to lighter, less restorative sleep.[1] The more sugar you eat, the more you crave, especially at night, which further disrupts your rest. Then you'll wake up feeling starved because poor sleep sends the hunger hormone ghrelin into overdrive. It's a vicious—and ravenous—cycle.

Sleep-promoting foods include high-fiber vegetables, which are delicious ways to reinforce your circadian rhythms. Kale and broccoli are great sources of tryptophan, a chemical that helps regulate melatonin. Mushrooms are full of vitamins D, B_2, and B_3, all essential as circadian reinforcers.

Pumpkin is a natural source of alpha carotene, which has been found to reduce sleep-onset insomnia.

Foods rich in magnesium help you maintain healthy levels of GABA, a neurotransmitter that promotes better sleep. Our bodies do not naturally produce magnesium, so it is essential that we get that vital sleep mineral in our diet through dark chocolate (limited to 30 grams a day), leafy greens, fatty fish like salmon and mackerel, quinoa, buckwheat, flax, chia and pumpkin seeds, legumes, nuts, and avocado.

Bananas, a high-glycemic fruit, are on Sarah's *don't* list. Lucky for us, the peel has three times the quantity of magnesium the fruit itself has. To get the magnesium and avoid the sugar, make my famous banana tea. Drink it an hour before bed to calm down, or anytime you feel stressed-out. Here's the recipe:

1. Thoroughly wash a banana to remove dirt, bacteria, and pesticides. Organic is best.
2. Cut ¼ inch off the top and bottom ends of the banana.
3. Leave the peel on and cut the banana in half horizontally.
4. Place the two banana halves in a saucepan with 700 milliliters of boiling water; let it boil for three minutes, until the banana turns brownish.
5. Strain the banana water into a cup.
6. Add a drop of honey or cinnamon if desired.
7. Drink.

WE'VE ALL GOT ENERGY TO BURN

The highest-quality energy source we have is fat.

If you set fire to a scoop of lard, it'd burn longer and slower than if you set fire to the same amount of sugar. Most of us are fortunate enough to carry around abundant supplies of fat in our own bodies, just waiting to be accessed for a nice, consistent, sustained energy burn. You might hate the fat on your belly or thighs (or both) right now. But it has a purpose: It is ready energy by the squishy handful, the best source of energy for you on the planet!

It's on point that our bulges are called things like "spare tires" and "saddlebags" because they are exactly that, places where we store energy for later use. Another charming phrase is "historical fat."

Historical fat doesn't have to be part of your personal story forever, though. To mobilize your body to use it, all you have to do is alternate periods of eating with periods of not eating, a method called intermittent fasting. This practice will help you squeeze decades-old fat cells dry and use that premium energy source to fuel your best life.

Science has discovered the easiest, fastest way to access that fat, and it's all about *timing*. An abundance of new research has found that *when* you eat is more important than *what* or *how much* you eat for energy.

During the eating period, or "window"—typically lasting eight, ten, or twelve hours—your body will use ingested carbohydrates to power itself and all it needs to do, like digest a meal, keep your blood flowing, grow muscle, regenerate cells, type at your desk, and power your brain.

During the fasting period, or "window"—typically sixteen, fourteen, or twelve hours—your body will use up any remaining carbs and, with no incoming fuel, start to use fat for energy. Depending on your metabolic speed and your weight, you'll shift into fat-burning mode at a slower or faster rate. For obese people, flipping the metabolic switch into fat-burn mode takes far longer than for lean people.[2]

Slow types might read this and think, "So I could fast for the same length of time as a Fast type, but I don't get as much fat burning as they do. *Not fair!*"

You're right. It's not fair. It just is. Accept it and use the knowledge of how *your* body works to inspire and motivate you.

LESS IS MORE

The whole concept of eating less for more energy seems counterintuitive, since we've been told our whole lives that food gives us energy. And it does! We need food to function. But intermittent fasting is not about starving yourself. You can eat the same number of calories, just over a shorter, specific period of time.

It's a mental shift, to wrap your mind around the idea of going twelve to sixteen hours straight (six to eight of those hours are while you're asleep) without taking a single bite. But if you regularly eat three meals a day and late-night snacks right before bed, all that food gives your body a surplus of carbs that is converted to fat and packed away in the saddlebags (not good). You aren't using your fat stores; you're adding to them.

It's not your fault, of course. Our bodies evolved to store fat, as much as possible. Back in caveman days, our ancestors often went days without eating and their fat storage kept them from starving to death. If not for our human talent to carry our own emergency fuel supplies on our bodies, our species would not have survived. The problem is, in modern times, with food abundance, we're only *adding* to our storage. Our body's capacity to pack on pounds is so efficient that if we were to eat nonstop, we'd get fatter and fatter until the excessive weight killed us.

Even if you ate only fruit, veggies, lean meat, and whole grain over too many hours, you'd still be on the metabolic pathway that breaks down carbs for energy rather than fat. Only once those carbs have been used up can you get on the fat-burning pathway. And the only way to do that is through *extended periods of fasting*. Remind yourself that fasting isn't starving yourself. It's giving your body the chance to use the food you already ate!

Intermittent fasting has proven benefits, along with fat burning. Fewer hours of eating means:

More ketones. Perhaps you've heard about high-protein, high-fat diets like keto and paleo, which are exceedingly restrictive about what you can eat. The objective of those diets is to shift the body's metabolism from being glucose-based (sugar-based) to being ketone-based. Ketones are the chemical byproducts made when the liver metabolizes fat for energy (what we all want). When you practice intermittent fasting for twelve hours or more, the ketone switch is flipped regardless of what you eat, according to a 2019 *New England Journal of Medicine* study.[3] So you can still have fruit or bread; just allow your body enough time to burn through the carbs and free up fatty acids to use for fuel.

More healthy cells. Studies have found that intermittent fasting increases stress resistance and longevity and decreases the incidence of cancer and obesity. Why would fasting help prevent cancer? Along with liberating fat to fuel your body, fasting activates the natural mechanism of "neuronal autophagy," or cleaning out old, damaged cells, aka "free radicals," and regenerating sparkling, shiny new ones.[4] Autophagy is a funny word that means whole-body repair and regeneration. Get to that healing state by fasting.

A faster metabolism. Fasting kicks your body's functioning into a higher gear. According to a recent Australian study, overweight women were put on an intermittent fasting schedule and high-fat diet for eight weeks.[5] At the end of the study, their energy expenditure — the number of calories used to maintain bodily functions, not counting exercise — increased. No wonder the participants lost weight. They burned more fat *just being alive.*

Reinforced circadian rhythms. The word "zeitgeber" means any external or internal force that cues the body's circadian rhythms to kick in. For example, the zeitgeber of darkness triggers the release of melatonin so we get sleepy. When we live in sync with our circadian rhythms, we function seamlessly. Fasting is a powerful zeitgeber, as recent research has discovered. For one study, scientists divided their mice subjects into two groups: The first group intermittently fasted; the second group ate whenever they liked.[6] Despite all the mice consuming the same number of calories, the eat-whenever group gained weight. The fasting group didn't gain weight, and it had a significant uptick in ketone-based metabolism, which led to robust activity in circadian rhythm genes. Under these biochemical conditions, the mice were more energetic during their awake periods and sleepier during their downtimes. When you eat at the same time of day every day, it's just like waking up at the same time of day. Consistency is the *key*. When your circadian rhythms are predictable, they're more efficient *and energized.*

Healthier choices. When people eat late at night, they're not reaching for carrots and celery. Researchers on the United Kingdom's National Diet and Nutrition Survey collected data on 1,177 adults over a six-year period about what and when they ate over the course of a day.[7] They found that, on average, 40 percent of the day's calories were consumed after 6:00 p.m. The participants who tended to stop eating earlier ate far fewer total calories. Per the food diaries kept by all the participants, the after–6:00 p.m. food itself was of the worst quality, high in salt, sugar, and empty carbs.

More good fat. We all know about the difference between good and bad cholesterol. High-density lipoprotein (HDL) does the good work of carrying cholesterol to the liver, where it can be filtered out of the body. Low-density lipoprotein (LDL) does the bad work of sticking to your arteries, clogging blood flow, and causing heart disease.

There are also two kinds of fat. "White adipose tissue" (WAT) is jiggly historical fat. We all have some and need it for insulation and to prevent starvation in times of scarcity. WAT becomes "bad" if the scale tips into obesity and increases our risk for all the associated illnesses, like diabetes, some types of cancer, and heart disease.

The other kind of fat is called "brown adipose tissue" (BAT). BAT is found in babies and hibernating animals, and in the muscles, shoulders, backs, and chests of adults. Its brown color comes from the density of nutrient-rich mitochondria that, when burned for energy, actually create heat inside the body in a process called thermogenesis.

WAT is a fine source of fuel, better than glucose. But BAT is the highest-octane fuel our bodies have to offer. When burned, it heats you from the inside, releases trace nutrients, increases energy expenditure, and decreases body weight. When humans are young, they've got BAT on tap. As we age, we have less and less of the good stuff and more WAT than we need.

After their eight weeks of intermittent fasting, the overweight women in the Australian study mentioned previously went through a process of fat browning called "recruitment." Some of their WAT turned into BAT. What this means: In terms of their body-fat composition and fat-burning ability, the participants got "younger and hotter" in just two months.

Do we need to say one more word about the benefits of intermittent fasting than that?

Everything in your Power Protocol—scheduling quality sleep, moving throughout the day, and intermittent fasting—promotes the BAT recruitment process.

Follow the protocol, and your body will change from the inside out.

Fasting allows your fat cells to burst with purpose into your bloodstream so they can power your muscles, brain, and other organs to do their jobs with four-bar energy. Use fat to hyperpower cell repair and regeneration, strengthen the collagen in your skin, and allow you to heal faster from injury. By turning WAT into BAT, your body will become a human torch, and the pounds will melt away.

> Energize Hack: Total-body energizing is available to anyone, and all we have to do is take our last bite of food early and not eat again until later the next day.

People hesitate when they hear the word "fast" and think they'll suffer extreme hunger after dinner. But there is no famine to fear. You won't feel pain. Studies have found that fasting increases levels of the "I'm full" satiety hormone leptin.[8] After a short adjustment period of about two days, your hunger hormone ghrelin will stop complaining[9] and your energy shift will speed up your metabolism, stabilize your blood sugar, decrease appetite, improve brain function, and boost your immune system.[10] You'll feel *up,* ready to *go* and *do.* Hunger is eradicated.

MICHAEL'S INTERMITTENT FASTING EXPERIMENT

I'm a Medium body type, but my weight crept up and up in my forties.

I decided to conduct some experiments with intermittent fasting on myself to see just how effective it is for weight loss, sleep, and, most of all, energy. We've all known that having a groaningly full stomach makes us sluggish. Overeating is an energy drain. Would eating for only eight hours a day be an energy gain?

I started with the simple plan of fasting for sixteen hours and having an eating window of eight hours, a 16:8 plan. It didn't seem too daunting. For at least six of my fasting hours, I'd be asleep. At first, I didn't think much about the specific hours I'd keep. I thought, *I get hungry around two in the afternoon, so I guess I'll start eating then and stop at ten at night.*

The first week, I dropped weight. My caloric intake was reduced as predicted. What I also noticed was that, though I was eating less, I felt less hungry. I'd been eating every three hours all day long out of habit my whole life. But when I stopped doing it, my body adjusted within days and those habitual alarms in my head that said, "You should snack now," just stopped ringing.

After three weeks, I experimented with food choice. Could I eat badly while fasting and still lose weight? For thirty days straight, I polished off an entire pint of ice cream every night within my eating window. I don't recommend this. All that sugar and dairy made me feel bloated and horrible. I did *not* feel energized by gas pains and having diarrhea every morning, but I sure did enjoy the ice cream while I was eating it! After a month, I checked the scale. I hadn't lost an ounce. But I hadn't gained an ounce either.

Through my own experimentation, I've figured out that my best eating window is 4:00 p.m. to 10:00 p.m. This is later than I'd recommend for most people, because I do eat after dark. Since I don't go to sleep until midnight or 1:00 a.m., I'm still going to bed on an empty stomach; this promotes quality sleep and reduces the risk of acid reflux along with other digestive problems. On this schedule, I'm in better shape. I've lost 20 pounds.

What eating pattern is best for you? We'll give you parameters based on chronotype and body type. But by tweaking the schedule via trial and error and thinking of your energy management as a fun experiment to conduct on yourself, you can fine-tune your schedule and get incredible results.

NEW RULES TO FAST BY

All our lives, we've been taught certain rules about eating and energy and have lived by them since childhood. Intermittent fasting proves some of those rules wrong, or right for the wrong reasons.

Old rule: Breakfast is the most important meal of the day.

New rule: Early eating has energetic benefits.

Breakfast has benefits, besides the yumminess of eggs on toast. For an English study of adults aged twenty-one to sixty, for six weeks, half the group was instructed to eat a substantial breakfast; half were told to fast until noon.[11] The study examined participants' energy balance based on three measures: (1) their resting metabolic rate, or how many calories each burned just by being alive, (2) their physical activity thermogenesis, or the energy they expended while exercising, and (3) their energy intake, or how much they ate.

The standout findings: (1) Among the leaner participants, eating breakfast caused an uptick in exercise thermogenesis, meaning Fast and Medium body types burned more calories via physical activity if they had a morning meal. (2) The breakfast eaters' blood sugar was more stable in the afternoon and evening than the non–breakfast eaters', a potentially lifesaving finding for Slow body types with a genetic predisposition for insulin resistance. If you're concerned about diabetes, breakfast is probably a good idea.*

Old rule: Dinner should be the biggest meal of the day.

New rule: Late eating causes problems.

Many of us grew up having a bowl of cereal for breakfast, a sandwich for lunch, and an ample plate of food plus dessert for dinner. The meal progression went from small to huge. But it's actually healthier and more energizing to do the reverse, or at the very least, avoid night eating syndrome, eating 25 percent or more of the day's total calories after dark. For an Israeli study, researchers split overweight and obese women into two

---

* If you have diabetes, intermittent fasting programs are not recommended. Consult your endocrinologist before making any changes to your nutrition and eating schedule.

groups and monitored their overall caloric intake.[12] For twelve weeks, the first group ate a large breakfast, medium-size lunch, and small dinner; the second group had a small breakfast, medium-size lunch, and large dinner. Both groups lost weight and reduced their waist circumference, but the early loaders lost more. Their glucose, insulin, triglycerides, and hunger hormone ghrelin levels were lower, too. The late loaders' triglycerides actually *increased* by 14 percent. Every health marker, as well as appetite control, was better for early loaders.

If you make only one change in your eating habits for huge energy gains, don't snack after your last meal of the day.

Old rule: For more energy, eat every three hours all day long.

New rule: Eat two meals and a snack on a consistent schedule.

Remember when "grazing" was a thing? The diet trend had us eating six or more small meals per day under the mistaken belief that consuming calories would stoke our metabolism. The opposite is true. Our metabolism is stuck in neutral when we eat. In fact, reducing meal frequency to two meals (or two meals with an interim snack) on a regular schedule, with an extended overnight fast between Meal 2 (the later meal, whether at traditional lunchtime or dinnertime) and Meal 1 (the early meal, whether at traditional breakfast time or lunchtime), reduces inflammation (the cause of cancers, diabetes, and heart disease), improves circadian rhythmicity (better sleep and daylight energy), boosts cellular repair and regeneration (disease prevention and antiaging), and enhances the function of your gut microbiome (less bloating and gas, and greater absorption of nutrients).[13]

SLOW METABOLISM AND INSULIN

For Slow types, carbs can wreak havoc on energy and health. Every time you eat carbs, they get converted into sugars. Your pancreas reacts to these dietary sugars by releasing insulin, which binds to receptors on blood cells and allows glucose to enter cells to give you energy. Along with sugar management, insulin also plays a role in fat metabolism. If you consume too much sugar over a long period of time, your pancreas will release more insulin to manage sugar levels, but the receptors that help deliver sugar to cells won't work as well. They're not as sensitive to insulin, and ultimately, they become resistant to it. Dietary sugar can't get into your blood cells, so it's converted into fat, which takes up long-term residence on your belly and hips. The more sugar you eat, the more insulin resistant you are, the more fat you gain.... The next stop on this train is diabetes, cancers, infertility, hypertension, and heart disease. It's not fair that your body craves the exact thing that can hurt you the most, but that is your unfortunate reality. It's better to know what's going on, accept it, and take small, simple steps to prevent the worst-case scenarios.

Slow types benefit from intermittent fasting more than Fast and Medium types for this reason alone. Fasting for sixteen hours, which we recommend for you, limits the release of insulin and makes your body more sensitive to it. In other words, when you do eat carbs after giving your pancreas a nice, long, sixteen-hour rest, your body will respond to insulin as it should, delivering sugars to cells for energy, speeding up your fat metabolism, and preventing diabetes. And you won't feel hungry.[14]

WHAT'S YOUR CIRCADIAN RHYTHM FASTING PLAN?

Following are three steps to calculate your ideal schedule and your energizing goals:

1. **Consult your doctor.** Intermittent fasting is not appropriate for everyone. You can't do it if you have an eating disorder (including binging issues), are diabetic, or are on insulin or some other medications. Check with your doctor before making any change in eating pattern. If you get clearance from your physician, proceed to the next step.

2. **Subtract three or four hours from bedtime to get your "last bite" time.** Sleeping with a full stomach causes digestive problems, disrupts sleep,[15] and has a negative effect on sleep quality.[16]

- **Lions:** Your bedtime is 10:00 p.m.; your last bite should be *no later* than 7:00 p.m.
- **Bears:** Your bedtime is 11:00 p.m.; your last bite should be *no later* than 8:00 p.m.
- **Wolves:** Your bedtime is 12:30 a.m.; your last bite should be *no later* than 8:30 p.m. But the earlier you stop eating, the better it is for your health.
- **Dolphins:** Your bedtime is 12:00 a.m.; your last bite should be *no later* than 8:00 p.m. But as you'll see in the chart that follows, we recommend that you stop eating even earlier.

A new study by researchers at Johns Hopkins University put male and female participants into two groups.[17] Both groups wore activity trackers and were rigorously assessed with body-fat scans and blood testing. One group ate dinner at 10:00 p.m. and the other ate the same meal at 6:00 p.m. Both went to bed at 11:00 p.m. After two weeks, the late eaters had gained weight and had *20 percent higher blood sugar levels* and 10 percent slower fat-burning ability compared to the early-eater group.

The early-riser Lions in the study who were forced to eat past their circadian bedtime did the worst of any subgroup. The late-riser Wolves were unaffected by the late dinner because they were already accustomed to eating—and digesting and metabolizing—later in the night than the Bears and Lions in the study. The only Power Profiles that can safely eat late consistently—past 9:00 p.m.—are Medium Wolves and Fast or Medium Dolphins. Everyone else is vulnerable to skyrocketing blood sugar and an elevated risk for life-threatening disease.

3. **Add the appropriate number of hours from your "last bite" to get your "first bite" time.**

The length of your fast depends on your metabolic speed.

Fast types: We recommend a 12:12 (twelve hours fasting; twelve-hour eating window) plan to reinforce your circadian rhythms and give you plenty of opportunities to fuel your already speedy metabolism.

Medium types: We recommend 14:10 (fourteen hours fasting; ten-hour eating window). The longer the fast, the more stored fat you will liberate.

Slow types: We recommend 16:8 (sixteen hours fasting; eight-hour eating window) to liberate plenty of stored fat, regulate appetite, lower blood sugar and cholesterol, and increase insulin sensitivity.

RECOMMENDED EATING WINDOW FOR EACH POWER PROFILE

Medium Bears	9:00 a.m. to 7:00 p.m.
Slow Bears	10:00 a.m. to 6:00 p.m.
Medium Wolves	10:30 a.m. to 8:30 p.m.
Slow Wolves	12:00 p.m. to 8:00 p.m.
Fast and Medium Lions	7:00 a.m. to 7:00 p.m.
Slow Lions	10:00 a.m. to 6:00 p.m.
Fast and Medium Dolphins	9:00 a.m. to 7:00 p.m.
Slow Dolphins	10:00 a.m. to 6:00 p.m.

What's your eating window? _____

STACEY'S INTERMITTENT-FASTING LIFESTYLE

I never loved to eat a big breakfast in the morning, even as a kid. I was a Snickers bar, Lucky Charms, or Cheerios kind of kid. My poor mom would set out the most perfectly cut apples and eggs, and all I remember doing was scrounging in the bottom of her handbag for change so I could buy something at the bus stop vending machine before school at 7:00 a.m.

The feeling of being hungry is manageable for me. I actually enjoy the lightness of my body when I'm not digesting food. I suppose it's a blessing that I have never struggled with food disorders like bulimia or had a warped perception of my own body. I have always had that athletic foundation that has carried me into my fifties quite well.

On the topic of "IFing"—a term I coined when I realized that intermittent fasting was the food "sport" I'd been playing—it wasn't intentional; it just happened. I think I just simply stopped eating a lot during a bad bout of canker sores in my mouth; the only time it was comfortable to eat was after I finished teaching spin classes around 11:30 a.m. or even as late as 1:00 p.m. on some days. If I had dinner the night before at around 6:00 p.m., that would mean I had gone close to sixteen or eighteen hours with no food without even realizing it. But I liked how I felt, and my abs were popping.

I brought this up with my very good friend and podcast partner, nutritionist maven Sarah Wragge, and she said, "You, my good buddy, are intermittent fasting on a 16:8 plan, and that's why you look so shredded!" Well, all righty! I'll just keep doing it, then!

It works for me, but it may not work for you, and if you are hangry and hate waiting around to eat so much that you want to hit the wall or throw your phone in the gutter, then I would maybe try to do a modified version of it.

It is completely my lifestyle now, but it doesn't own me. I choose to do it or not; most of the days I do, some I don't. On vacation, I will most likely blow it off, or make sure I do it every other day. I live my life, but intermittent fasting doesn't have control over my choice to eat. Ninety percent of the time, I eat clean, and 10 percent of the time, I drop a nice doughnut, a Coke, fries, and a hot dog (not at the same time!). Just promise me you'll stay in control of your food choices if you choose to fast, and don't become a jerk while you're doing it.

LIFE IN THE FASTING LANE

Once you get used to it and the energy gain it gives you, you'll hate going off your schedule. But in the meantime, smooth the transition with these golden nuggets of advice that worked for us.

⬆ **When you turn the lights on, eating is off.** Daylight hours are for eating; darkness is for fasting. Even Wolves like Michael who have their last meal later than other chronotypes can try to restrict eating after sundown.

⬆ **Hydrate.** As a general rule, hydration means 2 liters per day of water or herbal tea. Drinking will cut down on hunger cravings throughout the day (not only during your fasting period) and will improve the quality and duration of your sleep. Don't chug 500 milliliters of fluid within an hour of bedtime or you'll have to get up to pee.

↑ **Every bite counts.** We're not going to get preachy about sugar or demand that you eat more plants. Our concern is how you feel energetically about what and when you eat. So the only question to ask yourself on a regular basis is, "How will eating this food right now make me feel?" Experiment by eating something that will probably make you feel terrible and take note.

EATING COMPATIBILITY

We should make T-shirts that say, "Life in the Fasting Lane," to avoid the awkwardness of sitting down with someone at a restaurant or café during your fasting hours.

Eating compatibility can be a problem if your partner or friend group is used to having a late meal and your last bite is at 6:00 p.m. Of course, eating together is a joy and a bonding ritual the world over. And there is no reason your fasting schedule should interfere with that. Some solutions for working around our late-eating society:

• **Have a shared meal earlier in the day.** One of our Power Protocol test participants is a Slow Bear married to a Medium Wolf. Their late-meal timing is incompatible, but their early-meal timing aligns. Instead of planning a big shared evening meal—the custom many of us have kept—they make a point of breaking their fasts together every morning at 10:30 a.m. They're getting the same amount of bonding time, just on a new schedule.

• **Sit down anyway, and order tea.** Shared experiences don't require everyone to do exactly the same thing. You can sit down with someone to have a lovely catch-up, and enjoy herbal tea or decaf coffee (no milk or sugar, though, which would break the fast) while they enjoy their food. The objective is to be together.

- **Make it a point of health.** Intermittent fasting is for your health! No one would give you a hard time about not eating a particular food if you were allergic to it. If someone rolls their eyes at you for not eating at a particular time, simply say, "I'm concerned about my blood sugar and cholesterol, and this eating plan is my medicine," or "My weight is a concern, and eating on a schedule is making me healthier." And that is *that*.

TAKEAWAYS: EATING ENERGY

- High-quality fuel—plants (fruits, veggies, nuts, whole grains) and lean proteins (yes, meat is fine)—keeps the engine clean and prevents breakdowns. Sugar and grains clog our machine and weaken our energy system.
- The highest-octane fuel available is the fat you have stored in your body already.
- Access your fat for energy by limiting the number of eating hours per day, a method called intermittent fasting.
- Intermittent fasting has been proven to decrease inflammation, speed metabolism, reduce the risk of cancer and obesity, improve sleep, inspire healthier choices, decrease hunger, and convert "bad fat" into "good fat."
- The earlier you eat, the better for your health and resistance to disease.
- If your doctor gives you the green light to try intermittent fasting, you can start on a new schedule and begin changing your body from the inside out right away. Depending on your Power Profile, try twelve, fourteen, or sixteen hours of fasting between your last bite and your first bite the next day.

TO DO RIGHT NOW!

1. Take a look at your food choices and imagine ways you could cut back on sugar, dairy, and carbs in general. Begin to mindfully shift toward eating more lean proteins and plants.

2. Start adding up how much water you drink, and try to cut back on soda and energy drinks.

3. Notice how your calorie intake is balanced throughout the day. What percentage of your calories do you eat after dark? Think about how to reduce that number by a lot.

4. Consider doing more research on intermittent fasting. Go to TheSleepDoctor.com to learn more.

CHAPTER 6

Emotional Energy

When joy occurs naturally, it gives you an energy boost. But when Michael was getting his PhD in clinical psychology, he was taught that when people place a high value on happiness, they're more likely to feel disappointment, frustration, and anger at not reaching the level of exuberance they expected.[1] *Setting out to feel a certain way or placing high expectations on how we "should" feel is an energy drain.* We all want happiness, but self-prescribing it can have a paradoxical—and exhausting—effect.

Happiness can't be forced. Don't waste your energy on that pursuit.

However, going for "positive affect" is worth the effort.

⬆ **Positive affect** is a psychological term that means feeling good about how you engage with life, yourself, and your relationships. Prioritizing sleep, moving around as much as possible, and establishing an eating schedule are the top three items on a to-do list for positive affect. Packing your life and health with known sources of positive energy will make you feel stronger and better, both physically and emotionally. Being mindful about your health and energy is like plugging directly into an atomic reactor. You can *feel* those battery bars going up.

⬇ **Negative affect**—feeling bad about your interactions and engagement with life—can leave you flattened. We all know this to be true from our own lived experience of blaming ourselves when we're in conflict with a partner or colleagues, feeling like a failure, or hating ourselves for making a mistake. Now for some positive news about negative affect: *Science has found that feeling bad is energy draining only if we struggle against it.*

A Polish study looked at the intersection of emotional intelligence (understanding and responding to your own feelings and those of others) and chronorhythm. Predictably, researchers found that morning people were more energetically aroused (feeling *up*) early in the day, and night people hit their energetic peak after dark. Participants with the highest emotional intelligence (EI) rode their energetic waves. When they felt up, they did stuff. When they felt down, they rested. The lower EI group tried to power through downtimes and were more stressed and less energized because of it.

The researchers concluded, "These findings suggest that individual differences in circadian variation in mood reflect several factors, including an endogenous [internal] rhythm in energy, the distribution of social activities throughout the day, *and the person's awareness of their own energy level*"[2] (italics ours).

Just being aware that your emotions and energy ebb and flow over the course of the day can boost positive affect, energy level, and how you experience your feelings.

The secret to maintaining an energizing positive affect is simply to accept that emotions, like hormones, rise and fall naturally. That's just how we're built. No one can just stay chill all the time; it would actually not be healthy. Variation is important. By allowing our emotions to happen, we're not wasting psychic energy struggling against them or resenting them. Emotional energy is physical, on a chemical level. When the body releases dopamine and serotonin (happy, pleasure hormones) you feel good, and your mood and energy will lift. When the body releases stress hormones like cortisol, you'll feel tense and prickly.

Emotions aren't just tied to hormones, of course. Things happen, and we feel accordingly. Although we can't control everything that happens or how we react, we can choose to *accept* how we feel, and not squander energy by trying to change reality.

When we roll with our feelings, and live in sync with our chronorhythm, we gain energy and confidence in our ability to handle whatever life throws at us. The greatest source of energy you can ever tap into is being true to yourself, even when you're tense, mad, or sad. This is where your inner voice can either calm you or destroy you, so tune in to how self-talk can affect how you feel.

The real paradox of happiness: Allowing yourself to feel any and all emotions is more energizing than striving to feel just one, and it probably takes much less effort as well! Sounds like an energy gain for sure.

EMOTIONAL COMPATIBILITY

Each chronotype is known for certain habits, personality traits, and emotional profiles that might be a source of conflict. And conflict is itself an emotional drain. We're not saying that you shouldn't befriend or partner with someone with a different chronotype. Many psychological forces bring people together, and if someone fulfills an emotional need of yours, great! Only the strength of individual relationships can determine if, say, a neurotic Dolphin and rebellious Wolf can find love and friendship together. Any pairing can make it work. That said, it helps in a relationship to know as much about each other as possible—personality traits, the hours of the day when they feel good, and the hours of the day to leave well enough alone. With this information, you can better accept them for who they are, just as you'd like them to accept you for who you are. Knowledge is power and power is energy.

	Chronotype Traits	Happy Hours	Downtimes
Lion	Conscientious, driven, introverted, optimistic	6:00 a.m. to 3:00 p.m.	8:00 p.m. to 10:00 p.m.
Bear	Even-keel, friendly, extroverted, flexible	12:00 p.m. to 7:00 p.m.	7:00 a.m. to 11:00 a.m.
Wolf	Excitable, rebellious, risk-taking, moody	8:00 p.m. to 12:00 a.m.	8:00 a.m. to 3:00 p.m.
Dolphin	Neurotic, intelligent, loyal, obsessive	4:00 p.m. to 8:00 p.m.	6:30 a.m. to 2:00 p.m.

EMOTIONAL ENERGY GAINS

Along with acceptance and acknowledgment of our emotions, happy hours, and downtimes, positive affect is enhanced by *positive action*.

Committing to a sleep/move/eat schedule that is proven to boost emotions and energy is a profound example of taking positive action. We recommend a handful of other emotionally energizing actions that can add a bar or two to your body battery on command. Use them during your downtimes, or anytime you need a boost.

Laugh

Laughing is the best medicine—and a great power source. It boosts the immune system so your body doesn't have to work so hard or wear itself out fighting disease. It triggers the release of energizing dopamine, endorphins, and serotonin while reducing the fight-or-flight hormones cortisol and epinephrine. Stress hormones might give you a quick jolt of energy so you can run away from a figurative tiger, but once the chemical rush has passed, you're left drained.[3] Always allow yourself to feel certain things, but don't get stuck in any one emotion for too long. You'll just set yourself up for weirdness.

Watching *The Office* or *Zoolander* (never gets old) and laughing provide a mood lift, which reinforces other energizers like sleep. For an Indian study of elderly insomniacs in a nursing home, researchers divided the participants into two groups: One received "laughter therapy" and the other got "progressive muscle relaxation therapy" to see which modality would better aid sleep.[4] It turned out that laughter therapy was more effective than muscle relaxation and a lot more fun. *Comedians in Cars Getting Coffee,* anyone??

Not to pile on with the studies, but this one really amused us. For a Japanese study, nursing mothers with infants under six months were asked to watch at 8:00 p.m. either the Charlie Chaplin classic comedy *Modern Times* or a really boring hour-and-a-half-long show about weather.[5] Their breast milk was collected every two hours until 6:00 a.m. the following day, and those samples were tested for melatonin levels.

The moms who watched the Chaplin movie showed increased sleep hormone levels in their breast milk. If they hadn't had to pump breast milk all night long, they would have been able to fall asleep quicker and stay down longer.

> Energize Hack: Along with releasing energizing hormones and elevating mood, laughter before bed means better sleep and more energy the next day.

Listen to Music

Music gives us a direct hit of dopamine in the pleasure center of the brain[6] and reduces the flow of cortisol.[7] Listening with others — for example, at a concert or in a group therapy session — can relieve symptoms of depression and make people feel more confident and motivated.[8] When seniors listen to upbeat music, even in the background, their cognitive powers are supercharged. They process faster and have sharper memory functioning.[9]

You can use the energizing effect of music strategically to lift yourself up when you know you'll need a quick boost. You might do this already without being conscious of it. For instance, if you're a Bear, you might listen to '80s pop hits on SiriusXM while cooking dinner, right when you're in need of a second wind. If you're a Wolf, you might crank classic rock on your morning commute to shake off brain fog from lingering sleep inertia.

STACEY SAYS...

I *love* making playlists that include all eras! I curate music playlists for every activity in my life, and for my SoulCycle classes. My clients expect only the best from me, and when they go into the spin room and hear upbeat music, they get into a specific headspace before the ride based on the journey I've created for them. And that journey just continues to bring out the emotion for the next forty-five minutes. Music is the hidden magic of my group fitness classes, riding from past memory to future thought in just a short period of time.

Even when you are brought low and slow by depression or illness, listening to your favorite music — or any tunes, for that matter — can make you feel better. For a recent Brazilian study, researchers had middle-aged female radiology patients with breast or gynecological cancer participate in ten music therapy sessions. Despite the ravages of their disease, the participants reported feeling less fatigue, had fewer symptoms of depression, and had an improved quality of life in general (aka positive affect).[10] If listening to tunes can make cancer patients feel energized, imagine what it can do for you!

Be Kind

Kindness—in particular, being kind to yourself—is an energizing practice that can lift mood, drain tension, and help you sleep better. According to a comprehensive review of seventeen studies, across the board, the participants who had higher self-compassion (being accepting of one's faults and forgiving of failures, and avoiding self-critical thought) reported fewer sleep problems. Participants who were "self-cold" (shiver) had a harder time falling asleep and staying asleep for the night.[11]

Compassion toward others gives us a positive-affect boost, too. For a recent German-Swiss study, researchers hooked participants up to fMRI machines and had them watch a series of videos of people in distress and then describe their own feelings. This task lit up the parts of their brain associated with empathetic suffering. After a one-day compassion-training course where the participants learned meditation techniques and how to feel benevolence toward all people, their negative affect switched off and the brain regions associated with *positive* affect turned on. They saw the same images, but their brains reacted differently due to their new, nicer perspective.[12]

Small acts of kindness are enough to shift mood and generate energy. Even smiling at someone who seems down can lift your soul. Giving of yourself might seem like a way to expend energy, not make more of it. But perhaps opening your heart expends less energy than closing it. No data on that. Just our personal theory.

> ## STACEY SAYS...
>
> Alcoholics Anonymous encourages people to stay busy because they'll be more tempted to drink when they're doing nothing. They have a tenet called "be of service"—staying busy by helping people. When you're helping someone else, it takes your mind off your own boredom and off reminiscing about your partying days. Plus, it gives you a dopamine and serotonin rush, a feeling you might associate with alcohol and drugs.
>
> Helping, however you can, is an energizing tip that really works. It's one of my main motivators for doing charity work. I love helping other people and raising money for worthy causes. But I also know that when I give, I get back so much more.
>
> I believe that having a purpose is crucial to feeling energized. And for me, one of my purposes in life is to help others. It raises my energy and makes me feel good. It's love. It releases the love hormone oxytocin. I give of myself no matter what I do, but especially as a trainer. I bring people back to their highest energy potential, and doing that for others is the greatest power infuser in the world for me.

Soak Up Natural Light

We did a survey of a few hundred people on social media and asked, "What gives you a hit of energy when you need it most?" Most of the responses touched on common themes. Exercise. Call a friend. Play with a pet. Bake. Watch murder mysteries on BBC. Meditate. Do yoga. Read. Dance. Make jewelry.

The number one response by far was to take a walk or hike in the forest, at the beach, through a city park, along the river, with or without music or an audiobook.

Science backs up what your grandma always said: "Go play in the yard, you'll feel better." The outdoors is the most reliable and accessible emotional energizer there is. For a study published in the journal *Ecopsychology,* researchers split 181 participants into four groups: (1) indoor exercisers, (2) indoor resters, (3) outdoor exercisers, and (4) outdoor rest-ers. Their objective was to test changes in the participants' mood, energy, tiredness, and focus after they walked vs. sat, outside (in actual nature) vs. inside (in a room with simulated nature). Which variable was the most fatigue busting? Movement or environment? As it turned out, being outside was the most important factor. The participants who enjoyed the biggest jump in positive affect were the outdoor exercisers.[13]

If you have fifteen minutes and need to hit the reset button to shift negative energy (physical and emotional) toward the positive, taking a walk around the block will do more for you than relaxing inside.

Move More

We want to say, once again, that exercise is a mood booster, for all types. In a 2019 study, researchers at the University of Warsaw, in Poland, set out to pinpoint the intersection between mood, exercise timing, and chronotype among nearly 100 thirtysomething male and female CrossFit participants.[14] Regardless of when they worked out, all chronotypes reported significant improvement in three mood metrics: energetic arousal (feeling *up*), hedonic tone (feeling good), and tense arousal (feel-ing edgy and bad). **Regardless of chronotype, the subjects who showed up for the 6:30 a.m. sessions reported significant mood improvement in the post-exercise interviews.** (Sorry, Wolves.) Just doing the workout helped them overcome their natural aversion to being awake at that hour and raised all types to Lion-level enthusiasm. *No matter when you move, even if it's difficult, you will be thrilled you did.*

Moving and fitness participation tip the scales toward positivity for all energy types. According to a recent Australian study of more than 900 exercisers and 900 non-exercisers, researchers found a link between fitness, chronotype, and "positive personality-traits."[15] The personality traits tested were hope, optimism, perseverance, resilience, self-efficacy, and emotional intelligence. Turned out, sleep orientation wasn't nearly as indicative of positive personality traits as participation in sports.

Have Sex!

Sex can be a gratifying source of happiness and joy. Sleep deprivation inhibits the production of testosterone — the desire hormone — in both men and women, and it reduces sex drive and sexual responsiveness in women.[16] Being too tired to get it on means a lost opportunity to bond with your partner and bask in the glow of energizing closeness and connection and that powerful hit of oxytocin!

When is the most energizing time to have sex? Let's look at the science. In a recent Polish study of 565 participants aged eighteen to fifty-seven, researchers set out to determine when people felt the greatest need for sex, and when they were sexually active.[17] (See the charts that follow.)

Regardless of chronotype, female participants reported their time of greatest sexual need was 6:00 p.m. to midnight. The female Lions cited their second choice of 6:00 a.m. to 9:00 a.m.

Male participants were down with sex even during their least energetic time, if given the choice of sex or no sex. Male Wolves voted for 6:00 p.m. to 3:00 a.m. but would rouse their energy for a morning session at 9:00 a.m. to 12:00 p.m. Male Bears were down for it at 6:00 p.m. to midnight and 6:00 a.m. to 9:00 a.m. Male Lions voted for 6:00 a.m. to noon and were even willing to stay up until midnight if they had to.

As for when participants were actually having sex, across all chronotypes and genders, the most frequently reported time was 6:00 p.m. to midnight — that is, when the women reported their greatest need.

In the interest of upping sexual frequency, we've created these charts. But feel free to have sex anytime for more energy or any reason at all.

Heterosexual Couple	Male Dolphin	Male Lion	Male Bear	Male Wolf
Female Dolphin	8:00 p.m./8:00 a.m.	8:00 p.m./7:00 a.m.	10:00 p.m./8:00 a.m.	8:00 p.m./9:00 a.m.
Female Lion	7:00 p.m./7:00 a.m.	6:00 p.m./6:00 a.m.	8:00 p.m./7:00 a.m.	7:00 p.m./8:00 a.m.
Female Bear	8:00 p.m./7:30 a.m.	9:00 p.m./7:30 a.m.	10:00 p.m./7:30 a.m.	10:30 p.m./8:00 a.m.
Female Wolf	9:00 p.m./9:00 a.m.	9:00 p.m./9:00 a.m.	10:00 p.m./9:00 a.m.	11:00 p.m./1:00 a.m.

Male Gay Couple	Dolphin	Lion	Bear	Wolf
Dolphin	8:00 a.m./8:00 p.m.	7:00 a.m./8:00 p.m.	8:00 a.m./10:00 p.m.	9:00 a.m./10:00 p.m.
Lion	7:00 a.m./8:00 p.m.	6:00 a.m./6:00 p.m.	7:00 a.m./9:00 p.m.	9:00 a.m./9:00 p.m.
Bear	8:00 a.m./10:00 p.m.	7:00 a.m./9:00 p.m.	7:30 a.m./10:00 p.m.	10:00 a.m./11:00 p.m.
Wolf	9:00 a.m./10:00 p.m.	9:00 a.m./9:00 p.m.	10:00 a.m./11:00 p.m.	11:00 a.m./11:00 p.m.

Female Gay Couple	Dolphin	Lion	Bear	Wolf
Dolphin	8:00 p.m./9:00 p.m.	8:00 p.m./8:00 a.m.	9:00 p.m./10:00 p.m.	10:00 p.m./12:00 a.m.
Lion	8:00 p.m./8:00 a.m.	6:00 p.m./6:00 p.m.	9:00 p.m./7:00 a.m.	9:00 p.m./9:00 a.m.
Bear	9:00 p.m./10:00 p.m.	9:00 p.m./7:00 a.m.	7:30 p.m.	10:00 p.m./11:00 p.m.
Wolf	10:00 p.m./11:00 p.m.	9:00 p.m./9:00 a.m.	10:00 p.m.	11:00 p.m./1:00 a.m.

YOUR EMO-ENERGIZERS

Make a list of mood boosters that are guaranteed to energize you when you feel low and slow. Be specific! We know that laughter, music, kindness, and going outside will flood your tank with feelings fuel. But what makes *you* laugh? What Spotify playlist or SiriusXM station makes you feel like dancing? How can you be of service to someone else? What's your favorite green space to walk through? Be specific!

What gives *you* an energy surge? To take responsibility for your energy level, which is absolutely within your control to improve, you do have to know the emo-energizers that *you* can rely on. Maybe you love to dance to Duran Duran, play tug-of-war with your dog, have an afternoon orgasm. Write it all down and keep adding to your list. It might feel awkward to open a notes file on your phone and write, "Things That Give Me Life," but it's an essential part of acquainting yourself with your unique energy profile. And who gives a shit if what gives you energy is kind of awkward? Being an energized person feels a lot better than being a low-energy person. Give it a shot!

↑ **Michael's Emo-Energizers**

Ice cream

Helping people sleep

Great conversations
Laughing kids
Puppy breath
Babies yawning
Grand Slams
Underdogs winning
Hugging my kids
Teaching my kids
Driving fast with the top down
Really good movies
'80s and '90s rock
My dog
Making money
Winning

↑ **Stacey's Emo-Energizers**

Swimming in the ocean
Doing my job at 100 percent
Making someone smile
Hugs
Exercise
Buying plants and taking care of them
Taking photos with my iPhone
Paying bills on time
Dancing
Shopping for stylish unique clothes
Cooking a meal for family and friends
Watching '80s movies
Taking long walks
Helping friends out
Charity work
Taking my love on a well-planned date
Spending time with my family
Hanging with my friends

EMOTIONAL ENERGY DRAINS

We're all susceptible to psychic drains, joy suckers, buzzkills, negative-affect bringers that puncture our energy balloons. A major energy thief and mood downer is living out of sync with your chronorhythm.[18] Another is disappointment. When things don't go the way we anticipate, our energy shifts and joy dies. We can mitigate the negative effect of mood shifts by being aware of our vulnerabilities.

- **Lions** tend to be introverts and are exhausted by too much time in social situations. To recharge their batteries, they need to schedule ample alone time.
- **Bears,** on the other hand, gain energy in social situations. They need to schedule three social interactions per week to meet their recharge quota. Too much alone time drags their positive affect across the floor.
- **Wolves** are susceptible to energy-crushing mood disorders like anxiety, depression, and maladaptive behaviors like substance abuse.[19]
- **Dolphins** are at risk for anxiety disorders and depression, too. Brain-based disorders are not necessarily cured by following a Power Protocol, but being active and adequately rested goes a long way toward making people feel better and giving them the energy they need to manage their symptoms.

↓ **Michael's Joy Killers**

Melted ice cream
Physical or emotional pain
My kids in pain
Losing money
Anxiety
Losing a competition
Stupidity

Hard drugs

Nausea

Cold weather

↓ **Stacey's Joy Killers**

A bad attitude

Grumpy, rude, selfish, nosy, bulldozing, mean-hearted, know-it-all, judgy people

Bad weather on vacation

Traffic

An unclean car, house, or apartment

Passive-aggressive comments

Drivers who honk and don't let you merge

Spoiler alerters who ruin TV shows or movies

Parents who overcoach kids and make them cry

Not buying sneakers

Just as we asked you to be conscious of what lifts you up, it's important to know what drags you down, physically and emotionally. Maybe one depleting habit sets off a chain reaction of chronic tiredness and low motivation. Perhaps "eating ice cream at 10:00 p.m." or "talking to my mother when I'm already in a bad mood"? Make another list called "Things That Suck the Life Out of Me" and add to it as new data comes in. The simple act of writing down the activities that drain your energy will help you mindfully avoid them.

HOW SLEEP DEPRIVATION HARMS RELATIONSHIPS

Harmonious relationships are a big component of positive affect. *A major cause of conflict in otherwise emotionally uplifting relationships is sleep deprivation.*

Everything that makes us feel emotionally connected—and energized—is undermined by fatigue. Our ability to communicate, react rationally, and feel empathy declines. What happens when you're too tired and stressed to listen patiently and understand others' point of view? Anger, guilt, shame, resentment...every emo-exhauster on the spectrum. Engaging in bitter conflict with other people is like flushing positive-affect energy down the toilet.

What we lose in our personal relationships to exhaustion:

- **Emotional control.** For a recent study, researchers purposefully suppressed REM sleep in their 42 participants and observed their brain activity in an fMRI machine. The morning after, the participants' amygdala—the part of the brain that registers fight-or-flight emotions like fear—was lit up like a bowling alley.[20] Deprived of REM, your crankiness quotient goes way up. Life might be the same, but it *feels* scarier and meaner, and all day long, little stressors trigger overblown emotional reactions. An annoyance that, on a good-sleep day, wouldn't bother you at all can turn into an *outrage*. Instead of making smart, thoughtful decisions, your exhausted fear brain takes impulsive action that you're sure to regret later and have to waste precious energy fixing.

- **Emotional intelligence.** Lack of sleep makes you stupider about emotions. Our ability to read other people's facial expressions decreases.[21] Empathy is blocked.[22] We fail to interpret with accuracy emotional cues from friends, family, and colleagues, which can lead to draining fights, do-overs, and pain.

- **Optimism.** Fatigue focuses our brains on the negative. The more exhausted we are, the less capable of stopping the flood of "anticipatory anxiety," or the fear of bad things happening in the future.[23] A "Holy shit, we're doomed!" mindset—hardly an attractive quality in a long-term relationship prospect—is amplified by sleep deprivation.[24]

- **Well-being.** For a recent Turkish study, researchers set out to find the link between activity level, sleep quality, and well-being in 702 university students. First, they assessed movement level: Twenty-three percent were active, 57 percent were somewhat active, and 20 percent did a whole lot of sitting on their butts. Not too shockingly, the insomniacs (two-thirds of the total) were far more likely to describe their well-being as bad and to struggle with emotional problems and family drama. The active group had the highest psychological well-being scores, positive family relationships, and the lowest incidents of insomnia.[25] Which came first, activity level, sleep, or positive relationships? Does it matter? They nurture one another. You can have one without the others, but if you have them all, you have *it* all.

> ⚡ Energize Tip: To fill your energy tank with good feelings and peace of mind, and to ensure close, happy relationships, get more sleep.

STACEY SAYS...

I love socializing (as long as I can go to bed early). And I adore my close friends. But too much partying during a period of my life opened the door to alcohol and drug abuse. For me, addiction is the scariest and most devastating emo-exhauster of them all.

I've had my own journey with addiction. Others go through it with their own level of severity and horrible experiences. But none of them are good, and they're all destructive to your energy level and happiness.

Some major puzzle pieces have to come together to successfully overcome addiction. For some, it means rehab or support groups like AA or Narcotics Anonymous. I was fortunate enough that I never had to go to rehab; instead, I changed my geography and within two years, I had completely turned my life around. (You can read my story in *Two Turns from Zero*.) I know people who have been to rehab fifteen times and they still can't get it right. There're some really dark paths to go down, but I had to choose another way—to acknowledge my addiction and use the tools I had to change my energy and overcome it.

Many of my friends from that time in my life told me, "Come on, Stacey, everybody did drugs in the '90s. We weren't *that* out of control. We weren't *that* bad. Everyone was doing it." I lose energy listening to excuses and rationalizations. I know myself, and my addiction at that time held me back from a lot of success. I missed out on many professional opportunities because I was too busy partying.

When a close friend of mine committed suicide because of his addiction-related depression, I knew that my behavior had to stop. But it didn't, not right away. It took me a few months after he passed to get my shit together. And even then, I didn't change all the way until I moved from Los Angeles to New York. My New York friends were career-driven, with families and busy lives. They were not a bunch of crazy partiers.

Besides my new environment, what really changed my energy pattern, and ended my addiction once and for all, was exercise. When I started instructing spin classes at SoulCycle, I transferred my addiction from substances to teaching and working out. For me, it worked. I was lucky, and I know not everyone is. I used the power of prayer and meditation to shift my feelings and energy toward positivity. And I had a lot of personal angels—and angel power—to keep me on the right path. Here I am, fifteen years later, living a drug-free life that's so empowering—and energizing! I feel younger and stronger and more balanced than I ever have before. I'm going to be fifty-four soon, and this version of me could run circles around my former addicted self.

Drugs just will never be a part of my life again. I get a dose of clean energy every time I walk into a fitness studio or hang out with my tribe of people who are focused on exercise and wellness. Taking personal responsibility and holding myself accountable for my actions is how I energize every day.

TAKEAWAYS: EMOTIONAL ENERGY

- Nothing drains energy faster than trying to force a feeling. Instead of wasting energy, move with your natural, inevitable emotional ebbs and flows.

- Accepting your own emotional fluctuations, and the swings of others, helps you gain energy from relationships instead of draining it in conflict.

- To gain quick energy, do emotionally uplifting activities: laugh, listen to music, be kind to yourself and others, go outside, have sex.

- Be wary (and aware) of emotional drains, like sleep deprivation and substance abuse.

TO DO RIGHT NOW!

1. Start a list of your energy-gain activities by going about your life and noticing what gives you an emotional boost.
2. Start a list of energy-drain activities — including interactions with certain people — and mindfully avoid them.

Now you have a huge basis of knowledge about energy and exhaustion, and a very clear idea of what you need to do to increase your energy gains and decrease your energy drains. In Part II, each Power Profile gets a detailed Power Protocol of how to put all the science into practice. Read the chapter about you for you; read the chapters about the people you care about to better understand their energy issues and help them energize, too!

⚡

YOUR POWER
PROTOCOL

STOP AND DO

1. Before you read another word, **add a calendar entry** for *this Monday* that says, "Day One of My Energize Journey!"
2. **Program repeating alarms** for every action item on your Power Protocol. For example, on an iPhone, go to Alarm. Hit the + symbol. Enter a time (for example, 7:15 a.m.). Hit "repeat" for every day of the week. For "label," enter the action (for example, "Wake, Stretch, Drink"). Choose your alarm sound. Turn snooze "off." We know it's a bit of a chore (a tiny one), but it'll only take a few minutes, and then you'll be all set up to start energizing.

Success is a journey. The following is a four-week start-up plan to get you on your way and solidify it so it becomes a habit. Every week, you'll add more tools to your toolbox. By the end of the program, you will have all you need for an energized life. There's no rush to do everything right away. Take your time. Ease into it. The objective is to have more energy, not to drive yourself crazy or set yourself up for exhausting disappointment if you don't do it all exactly the right way immediately. Trying for perfection is stressful and defeats the purpose.

Each Power Protocol was created for your genetic chronotype and metabolism. The most important factor on this journey is understanding that your success depends on accepting and loving yourself. You can't change who you are, and why would you want to? If you schedule your sleeping, moving, and eating according to your DNA, you will have more energy, confidence, and power. We'll teach you how to overcome your energy drains and boost your energy gains. By working with your genes, you will achieve your energy goals.

Go time!

CHAPTER 7

The Medium Bear Power Protocol

Nina is a fifty-year-old chef in New York and a sporadic exerciser.* For six months, she hadn't been working out much at all. "I got a private catering job and with shopping, cooking, and transporting meals to my clients, I didn't have the two hours to get to the studio, do the class, shower and get back to the kitchen," she said. "I tried doing yoga at home, but I'd always find an excuse not to do it." As for movement during the day, Nina is on her feet during work hours. But as soon as the day is done, she would get flat on the couch.

Although weight gain has never been a big concern for Nina — she's been a size eight since college — she started to notice changes in her body. "Suddenly, I have cellulite on my legs and now my arm skin is doing this weird crepe thing. My muscle tone is basically gone, too. Used to be, I could fall off the exercise horse and my legs would look fine regardless. But now, if I don't work out for a few weeks, my whole body turns to mush."

Nina's social life revolves around food, since most of her friends and colleagues are also in the restaurant and catering business. She's constantly meeting people for meals at restaurants, going to dinner parties, and hosting her own. Even at home, she tastes and nibbles from morning until late at night. "I love food and wine. I *always* have tons of food in my freezer and fridge," she said. "My friends know that, if they come by, they will leave with lasagna or zucchini bread." She doesn't pig out per se. But she does graze on many small meals throughout the day.

---

* Names and identifying details have been changed. Some of the real-people stories in this book are composites.

She believed that falling off her exercise routine explained her exhaustion of late, but her sleep life wasn't helping. "During the last couple of years, I've gotten into this pattern of having a glass of wine at 9:00ish, binge-watching a show, dozing on the couch, and waking up an hour later, too tired to get off the couch and go to bed; so I'll doze for another hour. In the middle of the night, I'll crawl into bed, thinking, 'This is bad.' I wake up naturally around 7:30 and feel okay in the morning, but by noon, I'm beat."

What scared Nina: the idea that her energy was just going down from here. "Weary, worn-out, with an aching back and feet. I need to turn my energy around so that I can feel excited about the next ten or twenty years of my life. I hate feeling like I'm just getting by and that I'll get worse from here," she said.

Nina's energy goals:

- To sustain her energy into evening so she doesn't fall asleep on the couch.
- To get back in shape so she feels more powerful.
- To stop wasting energy feeling worried about her future.

WATCH OUT FOR THE BIGGEST ENERGY DRAINS FOR MEDIUM BEARS

Many of Nina's behaviors are sucking her energy dry.

⬇ Energy Drain: Grazing Is for Cows, Not Humans

The eating style of having many meals and snacks throughout the day is called "grazing." When you nibble every two or three hours from breakfast until right before bed, your body is always in "feed" mode and never gets the chance to go into "fat-burn" mode.

When you eat after dark, your blood sugar goes up when it should go down, and your inner clock spins in the wrong direction. On the one hand, your master clock knows it's dark out and that you should be winding down. On the other hand, incoming sugar and the insulin response

> **Bears,** your chronorhythm is synced with the sun. When it rises, your wake-up hormones—like cortisol and adrenaline—kick into high gear. Blood sugar, body temperature, and heart rate rise. When the sun sets, your settle-down hormones—like melatonin and serotonin—go up, and blood sugar, body temperature, and heart rate go down.

mean the day is *not* over and you are supposed to be awake. This push-pull causes sleep disruption, and then sleep deprivation, which, in turn, affects food cravings. The less sleep you get, the more your body craves high-fat, high-sugar foods. The more of that food you eat, the higher your blood sugar / insulin response, which makes your hunger increase. Michael wrote about these dangerous connections in *The Sleep Doctor's Diet Plan*. Sarah Wragge, the nutritionist, puts it succinctly: "Sugar begets sugar." Poor sleep and snacking late cause Nina's all-day grazing, and she's stuck in an exhaustion cycle.

↑ Energy Gain: Double Your Zeitgebers in One Shot

Sync up your sleep-wake circadian rhythm and your hunger-satiety rhythm by sticking with the intermittent fasting plan we've recommended for you. *Medium Bears, your feeding window opens at 9:00 a.m. and closes at 7:00 p.m.* Taking your last bite (or drink of anything but water or decaf tea) by 7:00 p.m. allows your digestive system to burn through the last remaining carbs by bedtime four hours later. During those fasting hours, your blood sugar will decrease, signaling to the master clock in your brain that the wind-down process of releasing melatonin can continue. When your body shifts into fat-burn mode overnight, your digestive rhythm will align with your sleep rhythm,

> **✳ Energize Hack:** To sleep better and make healthier food choices *effortlessly,* don't eat or drink anything other than water or decaf tea after 7:00 p.m.

and you won't have any cravings for the kinds of foods that turn hunger into an insatiable monster.

↓ Energy Drain: Not-Enough-Exercise Guilt

We've noticed this among Medium Bears—they have been active in the past and try to keep up with fitness routines. But they're also prone to letting things slide and feel guilty when they don't work out as often as they think they should or as much as they think their friends are. When they beat themselves up about being "lazy" compared to their peers, they're putting their health and lives at risk.

> ⚡ Energize Tip: In terms of our personal human potential, energy is not only the ability to change, but the essential ingredient we need to make change possible.

For a Stanford University study, researchers asked 61,000 respondents this key question: "Compared with others your age, are you physically more active, less active, or equally active?"[1] The findings were alarming: The participants who ranked themselves as "less active" than their peer group tended to have shorter lives than the ones who ranked their fitness as equal to or higher than that of their friends. In this perceived less-active group, some participants did, in fact, exercise enough. But the *perception,* the feeling of not doing enough, still shortened their longevity. Actual exercise didn't matter.

↑ Energy Gain: "You Go, Girl/Guy" Self-Talk: Leave the Guilt to Mom

To counteract negative perception, use positive, motivational self-talk. Just saying to yourself, "You're doing great!" cuts back on exhaustion and boosts energy. In a recent study with the intriguing title "Talking Yourself Out of Exhaustion," researchers asked people to cycle until tired.[2] Then, for two weeks, they gave themselves daily pep talks. After that, they came back in for another cycle session. Compared to

> ⚡ Energize Tip: Next time you feel bad about something, stop using the word "should," and start using the word "good." The self-talk shift alone will give you the energy boost you need to make positive changes.

the control group, the pep-talk group's rate of perception of effort dropped 50 percent and their cycling time-until-tired significantly increased. Just telling themselves they were doing fine allowed them to go longer and not feel like they were working harder.

⬇ **Energy Drain: Alcohol: Sleep Giver or Sleep Stealer?**

Drinking is fun, and Bears love a good time! There's nothing wrong with enjoying an adult beverage . . . as long as it's not outside of your eating window. Apart from the fasting issue, drinking an hour before bed doesn't give your body enough time to metabolize the alcohol, and that disrupts sleep.

Nina's pattern was to drink wine until she fell asleep on the couch, then wake up an hour later and doze again. Others might have to get up to use the bathroom a few times, which restarts the sleep cycle again and again, preventing the later restorative stages from occurring. It's like docking your phone and realizing hours later that it wasn't plugged in correctly. You get no energizing benefit despite all that time passing.

⬆ **Energy Gain: Cheers to a New Wine O'Clock!**

Limit drinking to two or three times a week, and *never* within three hours of bed or outside of your intermittent fasting schedule. A glass of white wine might be only 100 calories, but that's enough to break a fast. Limiting alcohol consumption can be a big change that you might not be too happy to make—at first. But after a week or two, you'll see steady improvement in your sleep quality and duration, weight loss from not drinking so many calories, and *increased energy* from the moment you wake up until you settle down for the night.

YOUR NEW KICK ASS ALL DAY, EVERY DAY PLANNER

Program these alarms on your phone for these key times:

7:00 a.m. Wake up; Stretch, plus water
9:00 a.m. Eating window opens
11:30 a.m. Shake, plus water

2:00 p.m. Power nap, twenty minutes or less (optional)

3:00 p.m. Bounce, plus water

5:00 p.m. Exercise (optional)

7:00 p.m. Eating window closes

8:00 p.m. Build, plus water

10:00 p.m. Power-Down Hour begins; turn off all screens

10:55 p.m. Balance (no water)

11:00 p.m. Bedtime

Week One

Begin on Monday. Any Monday. THIS Monday.

SLEEP ENERGY PRIORITY: RISE AND LEARN HOW TO SHINE

The objective for setting consistent wake times and bedtimes is to train your brain to follow a sleep rhythm that guarantees enough high-quality sleep and helps you to drift off quickly. Bears need a full night's rest, or they wake up groggy and grouchy. The first, most important step to that end is getting up at the same time each day.

- **Wake time is 7:00 a.m.**
- **Do not hit the snooze button.** *Repeat: Do not hit snooze!*
- **Get some sunshine.** Stepping outside or even just sticking your head out the window to get direct sunlight sends the rise-and-shine message along your optic nerve, straight into your master-clock brain center.

DAILY 5×5 PRIORITY: BOUNCE AND BUILD

Since Medium Bears are born with a degree of natural athleticism, your energy will increase quickly — almost instantly — from a combination of cardio and strength building. By the end of four weeks, you will do five daily movement sessions. But this week, to start off with a bang, Bounce at 3:00 p.m. Jumping will head off afternoon drowsiness by making your

149

heart beat faster. At 8:00 p.m., Build for muscle growth that'll make you feel strong and powerful. (See Chapter 4 for the exercises or go to MyEnergyQuiz.com for videos.)

- **Bounce.** At 3:00 p.m., do two minutes of Jumping Jacks followed by a one-minute rest. Then do two minutes of Jumps for Joy.
- **Build.** At 8:00 p.m., do one minute each of Squats, Crunches, Dips, and Wall Sits with twenty seconds of rest in between.

FASTING ENERGY PRIORITY: ESTABLISH AN INTERMITTENT FASTING RHYTHM

For you, we recommend the ten-hour eating window of 9:00 a.m. to 7:00 p.m. You get ten hours to eat however you'd like, keeping in mind the nutrition guidelines in Chapter 5. Grazers, get in the habit of circadian-reinforcing two scheduled meals per day with a snack in between, as long as your first bite does not occur before 9:00 a.m. and your last is no later than 7:00 p.m.

NINA'S WEEK ONE ENERGY SCORE: 3

"I found it really hard to stick with the eating and movement schedule, and that alone tired me out," she said. "But I didn't want to let myself down. I committed to a new wake-up routine and didn't allow for any snoozes. It felt like a sacrifice, but by the end of the week, I realized that I didn't need to snooze. By day three or four of sleeping through the night, my energy level definitely went up. I'm glad I stuck it out."

Week Two

SLEEP ENERGY PRIORITY: POWER-DOWN HOUR

Your melatonin starts flowing in the late evening, around 10:00 p.m. At that hour, begin a new bedtime routine devoted to quieting your sympathetic nervous system so you'll be able to drift off at or close to 11:00. We call it the Power-Down Hour. The point is to unplug and avoid the blue light from screens and to disengage your mind so you won't ruminate when trying to drift off. Your challenge will be sticking to that routine, which means getting off the couch and into bed for sleep (not Facebook) at 11:00 p.m. to complete roughly five ninety-minute sleep cycles you need for an energized next day.

- At 10:00 p.m., dock your phone, turn off the TV (or lower the brightness and make sure you're sitting ten feet from the screen), and shut down laptops.
- Instead of staring at a screen, listen to an audiobook or a podcast. Read a biography or light fiction. If you use an e-reader, put it on the dimmest setting or use blue-light-blocking glasses. Talk to your family (unless they are Energy Vampires). Do paperwork or an off-line non-work project.
- Take an Epsom salts and lavender bath ninety minutes to an hour before lights-out.
- If all your new energy is making you frisky, by all means, enjoy sex (alone or with someone else) during the Power-Down Hour and bask in the post-orgasmic glow of the love hormone oxytocin as it swaddles you gently in a cloud of positive energy. But plan accordingly. Get in bed earlier, so that you are ready for *sleep* at 11:00.

DAILY 5×5 PRIORITY: STRETCH AND SHAKE

This week, you'll start wake-up Stretches to limber up your spine for the day, and do a midmorning Shake to send oxygen-rich blood into stiff joints after sitting for hours. Continue Bouncing and Building.

- **Stretch.** At 7:00 a.m., do one minute of Child's Pose, one minute of Cat-Cows, one minute of Sphinx, one minute of Dragonfly, and one minute of Cannonball. These five poses will become your daily morning flow.
- **Shake.** Do a full circuit of Shake at 11:30 a.m. to get your blood circulating. One minute each of Neck Looseners, Arm Circles, Leg

Swings, Crescent Bends, and Trunk Twists, with a few seconds of rest in between. This will become your daily midmorning flow.

- **Bounce.** For an afternoon energy jolt at 3:00 p.m., Skip up and down the block for two two-minute sets with one minute of walking in between. If you're stuck inside, Skip in place.
- **Build.** Along with a circuit of Squats, Crunches, Dips, and Wall Sits (one minute each with a few seconds of rest in between), add one minute of Kicks at 8:00 p.m.

FASTING ENERGY PRIORITY: TRACK SNACKING

Bears love to nibble and often eat out of boredom, habit, or opportunity (the plate of cookies is just sitting there; might as well have a few) when they're not actually hungry. This week, though, be mindful of the quantity of food you eat and when you feel real hunger. Along with committing to your intermittent fasting plan:

- When you first feel a pang of hunger, have a glass of water and wait twenty minutes. If the hunger is real, the feeling won't go away.
- Listen to "I'm full" signals from your gut. Leptin is the satiety hormone, and it becomes desensitized by grazing. When you shift to a meals schedule, the signal to stop eating will gradually become louder. Listen for it, and then stop. You'll wind up eating a lot less.

NINA'S WEEK TWO ENERGY SCORE: 4 (+1 FROM THE PREVIOUS WEEK)

"I really missed my TV after 10:00 p.m. and couldn't quite turn it off. So I turned the brightness way down. I hope that made a difference. I think I did maybe two movement sessions per day even though I was supposed to do four. I thought that, since I'm on my feet in the morning, I didn't need to Shake. But that was wrong. My body was upright but still stiff. By shaking out my arms and legs, I felt more energy. The back and neck moves have eased my chronic pain from always looking down. Not feeling achy all day is a real energy boost. I'm upping my game next week."

Week Three

SLEEP ENERGY PRIORITY: GET OUT OF YOUR HEAD!

When life gets complicated, Medium Bears get anxious and tend to ruminate, or think about problems (sometimes obsessively) they can't necessarily control. Ruminating at bedtime is the leading cause of insomnia for most humans, including Bears. To avoid anxiety, you'll reach for your melatonin-inhibiting devices for distraction, wind up watching TV too late, and then get stuck in a pattern of sleep deprivation. Being crushed under the burden of sleep debt is exhausting, especially for Bears, who really need seven to eight hours per night.

We recommend a two- or three-minute session of diaphragmatic breathing—deep belly breaths—during your Power-Down Hour to help prepare your body for sleep and block rumination before it begins. Here's how you can do it:

↑ Calming Diaphragmatic Breathing for Beginners

1. Sit in a comfortable chair or lie flat on the floor on a mat.
2. Place one hand on your belly just beneath your ribs. Place the other hand on your sternum.
3. Inhale deeply through the nose into your belly until your hand rises for a count of four. Your chest should remain still.
4. Hold for four.
5. Exhale through pursed lips and let your belly and hand deflate for four. Keep exhaling until all the air is expelled.
6. Repeat ten times. Keep it slow and steady.

DAILY 5×5 PRIORITY: BALANCE

For your final movement session of the day, Balance at 10:55 p.m. is just one more way to quiet the mind and prepare yourself for sleep. Incorporate Balance into your Power-Down Hour.

- **Stretch.** At 7:00 a.m., do the morning flow of spine Stretches you know by heart by now.
- **Shake.** At 11:30 a.m., run through your circulation Shake movements.
- **Bounce.** Time to take it up a notch at 3:00 p.m. and try more challenging Bounces: Ice Skater Jumps and Burpees. Two minutes of each, with a short rest in between.
- **Build.** At 8:00 p.m., it's time for your circuit of Squats, Crunches, Dips, Wall Sits, and Kicks. One minute per movement with a few seconds of rest in between.
- **Balance.** At 10:55 p.m., make like a Tree for five minutes, alternating feet every thirty seconds.

EMOTIONAL ENERGY PRIORITIES: DON'T BE A DOORMAT—YOUR KINDNESS CUTOFF

What often happens with friendly Bears is that they give to others and expend their own energy in the process. It feels good to give, as we've noted. But if you give too much of yourself, there might not be much left over for you. Energy means the ability to *go* and *do*. But if givers do too much for takers, they pay a price.

> ⚡ Energize Tip: Set a kindness cutoff point. Do one or two favors per day and soak up the energizing lift. But defer further favors until the next day. For chronic givers, learning to say, "I'm sorry but I can't today," can boost their energy levels.

Nina spent much of her week doing things for friends, family, clients, and colleagues. She was asked, "Can you do me a favor?" to exhaustion. The phenomenon of being the person who does for others in the workplace is called "organizational citizenship behavior." One study of 273 employees linked the behavior to "citizenship fatigue"—feeling tired, worn-out, taken advantage of, and unsupported.[3]

NINA'S WEEK THREE ENERGY SCORE: 5.2 (+1.2 FROM THE PREVIOUS WEEK)

"Okay, let's talk about the wine. I cut back to two glasses of wine this week. I missed my nightly glass (or two). But, just like TV, I realized after a few days without it that I didn't need it, and that the benefits of sleeping well far exceeded the buzz of drinking."

Week Four

SLEEP ENERGY PRIORITY: THERE'S A NAP FOR THAT!

Feeling tired in the afternoon is *normal* and *natural* for Medium Bears. Acknowledging your chronorhythm is energizing in and of itself. Heeding it is the key to your personal power. So, on those drowsy days when Bounce won't be enough to revitalize you after lunch, take a power nap.

For Bears, a short nap (no more than twenty minutes) will restore you to morning levels of energy and alertness and give you a productive afternoon. If you go over twenty minutes, though, you'll come out of a nap groggy and stay that way for an hour. Try Michael's Nap A Latte technique (see page 69), or, if you'd rather, just do some diaphragmatic breathing in a darkened room or go outside and do a few extra Bounce exercises instead of napping.

> **Energize Hack:** The best time to nap is approximately seven hours after waking, so if you wake at 7:00 a.m., nap at 2:00 p.m.

DAILY 5×5 ENERGY PRIORITY: GO HARDER

This week, up the intensity.

- **Experiment.** Are there any movements you have been afraid of trying, like Eagle, Dancer, or Gorilla Burpees? This is the week to face your fears and experiment with the movements that you've avoided.
- **Intensity.** Level up. If you have been coasting with Easy level, go Medium. If you've been doing Medium, go Hard. If you've been doing Hard, up the amplitude. Jump higher, Skip faster, Squat deeper.

- **Duration.** Extend your Daily 5×5 to a Daily 5×7. Add two min-utes to each session if you are feeling it!
- **Frequency.** If you've only been doing three sessions a day—some days are just like that—really push this week to hit the magic number five.

FITNESS FOR MEDIUM BEARS

You were born naturally athletic. Perhaps, like many Medium Bears, you have been in excellent shape at different times in your life, but for whatever reason, you haven't maintained your fitness. To reclaim the body you once had, participate in fitness activities at two well-timed opportunities:

- **7:30 a.m. to 8:00 a.m. strength building.** After stretching, add fif-teen to thirty minutes of strength training. The endgame is to gain muscle and lose fat. Medium Bears can gain muscle and lose fat pretty easily, but only if they train. A fifteen-minute burst of lunges, squats, planks, and push-ups—anything that makes your muscle fibers twitch—will do the trick. Why first thing? Morning movement is a setup for all-day energy. If you can do it outdoors and get a morning dose of sunlight, it'll strengthen your circadian rhythms, too.
- **For an extra energy boost, try this in your *bare feet,* on the *earth*** (yes, touch the ground with your toes). While the data on "ground-ing" is not quite there yet, we both seem to get a boost from it.
- **5:00 p.m. cardio.** Bears are at physical peak at 5:00 p.m., with maxi-mum lung capacity and hand-eye coordination. We recommend group fitness (in person or online) for you because you gain energy from the enthusiasm of those around you. Try indoor cycling, a strength-training class, or a high-intensity interval training (HIIT) class. Or just take a forty-five-minute jog or power walk with a friend.
- **Cardio can send your Medium metabolism into overdrive.** Pro-vided that you follow the prescribed Power Protocol, you won't be too exhausted at the end of the workday to attend a fitness class or go for a run, and it won't feel nearly as painful as it would at any other time of day. Your "perceived exertion," or how hard you think you're working, is at an all-day low in the early evening.[4] As a result, you'll push yourself harder, which is great for endomorphs, who need all the motivation they can get.

FASTING ENERGY PRIORITY: SYSTEMS CHECK

After three weeks of intermittent fasting, you have already adapted to the change in your eating rhythm and have probably noticed that your pants are a bit looser or that your muscles have new definition. This week, see if you can push your window by thirty minutes of additional fasting. If you just *can't*, okay. But you won't know unless you try.

- **Open** your eating window at 9:15 a.m.
- **Close** it at 6:45 p.m.

NINA'S WEEK FOUR ENERGY SCORE: 6 (+3 SINCE WEEK ONE)

Nina started this program with a few energy goals in mind:

- **To sustain her energy into evening so she doesn't fall asleep on the couch.** "I felt like it all came together at the end. I was averaging around four daily movement sessions, which was four more than I did before! Cumulatively, better sleep, not drinking after dinner, and moving more has made me feel a lot better physically and emotionally. I can see the direction I'm headed, and it's toward energy, away from exhaustion."
- **To get back in shape so she feels more powerful.** "I didn't find time for fitness this month, but unlike before, I'm not going to beat myself up about it. I found that I felt so much better sleeping, eating, and moving on a schedule that I didn't really need to work out as well. I felt energized without it, so why cram one more thing into my day? I don't need the stress or guilt! I'm doing enough. I am enough."
- **To stop wasting energy feeling worried about her future.** "This process boosts energy from day one, but it is cumulative over time. The more you do, the more you *can* do. The farther you go with it, the farther you will go in life. So if you feel discouraged that you can't do it all, all at once, be kind to yourself. Do what you can, as you can, and you'll reap the rewards. And that is the attitude that I can live with, and thrive with, for the next thirty years of my life!"

Nina's energy score nearly doubled in one month's time, and she was thrilled by the improvements in every aspect of her life. To increase her energy as she moves forward, all she needs to do is stay mindful about her goals and be kind to herself.

Another life energized. Our work here is done.

TROUBLESHOOTING

Medium Bears, a couple of things to watch out for:

1. **Frustration.** Some of the movements will feel awkward until you practice them enough. Ice Skater Jumps are hard to master. But who cares if you feel like a goof? Perfection is not the point. Silence the Olympic judges in your mind and give yourself top marks for effort.

2. **Backsliding.** After a few good days or weeks, you might think, "I've earned a break." Exhausting yourself with overeating and being sedentary is no reward; it's a self-inflicted punishment. And then one "off" day rolls into another, and you're right back where you started. Cheat days only cheat you. Fight backsliding by meeting it with energizing action. Do something that will lift you up (sex, or dancing, or cooking a fantastic meal with friends).

Daily Schedule

The end of Week Four is the beginning of the rest of your high-energy life. Stick with it, and you'll always feel the power.

7:00 a.m. Wake up; 1 minute Child's Pose; 1 minute Cat-Cows; 1 minute Sphinx; 1 minute Dragonfly; 1 minute Cannonball; drink 500 milliliters of water

9:15 a.m. Eating window opens

11:30 a.m. Level up movements — double time optional: 1 minute Neck Looseners; a few seconds rest; 1 minute Arm Circles; a few seconds rest; 1 minute Leg Swings; a few seconds rest; 1 minute Crescent Bends; a few seconds rest; 1 minute Trunk Twists; drink 250 milliliters of water

2:00 p.m. Power nap (optional)

3:00 p.m. 5 minutes Medium or Hard jumps of your choice; drink 250 milliliters of water

5:00 p.m. Exercise (optional)

6:45 p.m. Eating window closes

8:00 p.m. Level up movements — double time optional: 1 minute Squats; a few seconds rest; 1 minute Crunches; a few seconds rest; 1 minute Wall Sits; a few seconds rest; 1 minute Dips; a few seconds rest; 1 minute Kicks; drink 250 milliliters of water

10:00 p.m. Power-Down Hour; turn off all screens

10:55 p.m. 2 minutes Dancer; 2 minutes Eagle; 1 minute Figure Four

11:00 p.m. Bedtime

CHAPTER 8

The Slow Bear Power Protocol

Julia, a forty-five-year-old management consultant and mom in New York, read a blog on the Sleep Doctor site about the link between sleep deprivation and weight gain and thought, *That's me!* "I put on extra pounds so easily," she said. "If I just think about a piece of cake, it appears on my hips the next day," she explained. "When I hit my forties, my weight became a serious concern. I'm about 25 pounds heavier now than I was at thirty. If I keep up this pace, I'll be 200 pounds by sixty. It's not healthy. I feel bad about myself when I look in the mirror. My doctor told me that my blood sugar is on the verge of being prediabetic, which is frankly terrifying. If getting more sleep will help me lose weight, I'd be thrilled. But that's easier said than done."

Her sedentary, stressful lifestyle isn't helping. "I'd describe my day as a constant battle to stay on top of everything I have to do for my family and my job," she said. "My twin daughters are thirteen, and they depend on me to organize every detail of their lives. I've got all the household shopping and maintenance to deal with on top of my full-time job. There isn't enough time for exercise or eight hours of sleep. We barely see our friends, either. I'm a social animal, and when I feel isolated, I start eating cookies. Don't ask me the last time I chopped and cooked vegetables. It's too embarrassing to admit."

When she finally goes to bed, Julia scrolls through Apple News, Facebook, and Twitter. "It's a ritual that has become a habit I can't seem to break. I might be on my phone until midnight or later," she said. "My husband grumbles at me to turn it off, but it's hard to switch off my device *and* my brain. I wind up sleeping six hours a night on weeknights.

Sundays are my designated morning for 'me time,' when my husband gets the kids out of the house. I always use the time to catch up on sleep and don't get out of bed until noon or one. But that sets me up for deep trouble on Sunday nights."

Julia knows that more sleep and movement throughout the day would go a long way toward giving her more energy. "I'm not an idiot!" she said. "Everyone knows that. But I've got no energy to change my habits. I feel trapped inside my exhaustion."

We asked Julia for her energy goals and she said, "I don't even have the energy to really think these through." But she said her goals were to:

- Lose weight.
- Sleep better.
- Move more.

WATCH OUT FOR THE BIGGEST ENERGY DRAINS FOR SLOW BEARS

Many of Julia's behaviors are sucking her energy dry.

↓ Energy Drain: Sleeping Late on Sunday, aka Social Jet Lag

As we've discussed, sleeping late on Sunday can lead to Sunday night insomnia, which is a particular challenge for Bears, who really need seven to eight hours to function. Instead, on Mondays (and probably Tuesdays and Wednesdays as well), they'll wake up after five or six hours grouchy and sluggish, which makes them crave high-sugar, quick-energy foods. Sleep deprivation raises Slow Bears' risks for weight gain, mood disorders, life dissatisfaction, obesity, diabetes, and heart disease.

↑ Energy Gain: Sleep Better During the Week

Do the energy math: If you only get six hours of sleep Monday through Friday, you are in "sleep debt" of two hours per night, times five nights. Ten hours of sleep debt cannot be recouped over the weekend.

The data shows all you do is get more sleep-deprived as the weeks go on. We're prescribing a pre-bed routine that'll help you get more sleep all week, so your Sunday "me time" can be spent doing something that gives you energy, like brunch with friends. If you absolutely *must* sleep in on the weekends, we have a hack for you: On Saturday and Sunday mornings, don't sleep later than forty-five minutes beyond your weekday wake time. That won't send you into social jet lag.

↓ Energy Drain: The Couch Is a Human Magnet

Newton's law says, "An object at rest stays at rest." Slow Bears, more than any other type, are most likely to park themselves and not want to move again for hours. If you're cozy in bed, you may hit the snooze button twice before you can get up. It'll take a grenade to get you off the couch once the lounge pants and Netflix go on. When you are plonked down on the furniture, you're not getting outside much. Julia's life went from home to car to work to home on repeat. And when she was at home, her movement was limited to walking from the kitchen to the living room to the bedroom. She said, "I do laps on a track: from the couch to the fridge, over and over. At night, I hit that every hour." The sedentary lifestyle goes hand in hand with overeating—and the weight gain and insulin insensitivity that Julia is so worried about.

↑ Energy Gain: Repel the Magnetic Force of Doing Nothing

Newton's law also says, "An object in motion stays in motion."

> **Slow Bears Energize Hack:** Set alarms on your phone for sunshine breaks. Just five minutes of direct sunlight in the morning—ideally within thirty minutes of waking up—and another fifteen when your cortisol dips at 2:00 p.m. are enough to recharge your master clock and give you an energy jolt.

When Slow Bears feel the gravitational pull of inactivity and the predictable siren call of sitting, the most energizing thing you can do *right away* is to go outside and get some energizing vitamin D. Sunlight is a power zeitgeber. It comes in through your eyeballs, travels along the optic nerve, and passes straight into your circadian command center.

↓ Energy Drain: Snacking After Dinner, or "What's Wrong with a Little Popcorn?"

Julia's top concern is her steady weight gain, and eating late at night is causing it. Slow Bears: It takes your body a long time—four to eight hours—to use up the calories you eat before shifting into fat-burning mode. So if you are continually snacking, you rob yourself of the opportunity to make that switch and lose fat that's weighing you down and robbing you of energy.

↑ Energy Gain: Don't Eat After Dinner

This is where your self-control *must* kick into high gear. When you commit to the intermittent fasting plan we recommend for Slow types—an eight-hour eating window, from 10:00 a.m. to 6:00 p.m., followed by sixteen hours of fasting—your body will have eight to twelve hours to burn fat, slow insulin, improve sleep, and increase energy.

Having an early dinner and then not eating another bite until breakfast is a major change for anyone who is used to snacking at night, and it's going to be hard—at first. Your stomach will rumble. Your emotional brain center, the amygdala, is going to scream, *"Feed me!"*

The pain won't last. It takes approximately three days to adjust. But by the end of Week One, you will be astonished that late-night cravings are just . . . gone. Your gut-based circadian clock will reset itself. From then on, snacking at night will feel wrong and uncomfortable.

> ⚡ Energize Tip: Late-night hunger is an emotional habit. It has nothing to do with needing nutrients to fuel your body. If you are hungry after dinner, drink water or hot tea.

YOUR NEW KICK ASS ALL DAY, EVERY DAY PLANNER

Program these alarms on your phone for these key times:

7:00 a.m. Wake up; Stretch, plus water
10:00 a.m. Eating window opens

11:30 a.m. Shake, plus water

2:00 p.m. Power nap, twenty minutes or less (optional)

3:00 p.m. Bounce, plus water

5:00 p.m. Exercise (optional)

6:00 p.m. Eating window closes

8:00 p.m. Build, plus water

10:00 p.m. Power-Down Hour begins; turn off all screens

10:55 p.m. Balance (no water)

11:00 p.m. Bedtime

Week One

Begin on Monday. Any Monday. THIS Monday.

SLEEP ENERGY PRIORITY: RISE AND LEARN HOW TO SHINE

The objective for setting consistent wake times and bedtimes is to train your brain to follow a sleep rhythm that guarantees enough high-quality sleep and helps you to drift off quickly. Bears need a full night's rest, or they wake up groggy and grouchy. The first, most important step to that end is getting up at the same time each day.

- **Wake time is 7:00 a.m.**
- **Do not hit the snooze button.** *Repeat: Do not hit snooze!*
- **Get some sunshine.** Stepping outside or even just sticking your head out the window to get some direct sunlight sends the rise-and-shine message along your optic nerve, straight into your master-clock brain center.

DAILY 5×5 PRIORITY: BOUNCE AND SHAKE

Your biggest challenge is to move regularly throughout the day. Unlike fidgety Fast types, who don't stop twitching, and Medium types, who take the stairs instead of the escalator just to show off, Slow chronotypes need a lot of motivation to get moving. But once you get in motion, you're more likely to stay that way, and see some really cool, fun places you've always wanted to visit (like the Land of the Slimmer). Just standing up and walking around every hour will boost mental clarity and stamina. It doesn't take much to notice fast change as long as you move consistently throughout the day. By the end of four weeks, you will do five daily movement sessions. But this week, start with a *bang* and begin your daily practice of Bounce and Shake movements. (See Chapter 4 for the exercises or go to MyEnergyQuiz.com for videos.)

• **Shake.** At 11:30 a.m., after a few hours of sitting, it's time to stand up and get your blood moving to your cramped muscles and stiff joints. By doing a five-movement flow, you'll loosen up every part of your body. Do one minute each of Neck Looseners, Arm Circles, Leg Swings, Crescent Bends, and Trunk Twists, with a few seconds of rest in between.

• **Bounce.** At 3:00 p.m., do two minutes of Jumping Jacks followed by a one-minute rest. Then do two minutes of Jumps for Joy.

STACEY SAYS...

Bounce and Shake not only loosen joints, increase circulation, and give you energy; they also relieve tension. We've recommended the perfect time for these moves (11:30 a.m. for Shake and 3:00 p.m. for Bounce), but you can break them out whenever you feel that first dose of anxiety over all you have to do. If you don't have privacy at work or at home, find an empty office or even go into a bathroom stall to move and relieve stress. (Don't laugh: I've done Jumps for Joy in many a bathroom over the years!) When you resume your day, you'll feel powerful, energized, like you can *crush* it. While moving, always remember to take deep breaths. Gabrielle Reece told me, "You know we're getting fit just by breathing. Breathe in energy and power. Breathe and believe."

FASTING ENERGY PRIORITY: ESTABLISH AN INTERMITTENT FASTING RHYTHM

For you, we recommend an eight-hour eating window of 10:00 a.m. to 6:00 p.m. Only eight hours to eat, no more. Grazers, get in the habit of circadian-reinforcing two scheduled meals per day with a snack in between, as long as your first bite does not occur before 10:00 a.m. and your last is no later than 6:00 p.m.

STACEY SAYS...

Some of you may be wondering, "I have work dinners that start at 7:30 p.m. How am I going out to a dinner and not eating?"

Answer: Order the healthy option, skip the appetizers, and take a couple of bites, leaving the rest to take home with you. Let the table know you are on a mission to stick with your new health plan, and call it a night. If this isn't possible, push your next day's breakfast off until lunchtime. People will be impressed by your commitment to health, but it can hold a mirror up to someone else's horrible eating patterns. Be strong, and be ready for the backlash. People can say the most messed up things when you are the one in control of your food!

JULIA'S WEEK ONE ENERGY SCORE: 3.5

"I have to say, this worked like magic, and very quickly. After about six days, I didn't need to set my alarm," she said. "I automatically woke up at the same time. Until my body adjusted to my sleep schedule, I got around six hours per night, not enough for me. My week started with pretty low energy ratings, but by the end, when I conked out on schedule, my energy rating went up."

Week Two

SLEEP ENERGY PRIORITY: POWER-DOWN HOUR

Your melatonin starts flowing in the late evening, around 10:00 p.m. At that hour, begin a new bedtime routine devoted to quieting your sympathetic nervous system so you'll be able to drift off at or close to 11:00. We call it the Power-Down Hour. The point is to unplug and avoid the blue light from screens and to disengage your mind, so you won't ruminate when trying to drift off. Your challenge will be sticking to that routine, which means getting off the couch and into bed for sleep (not Facebook) at 11:00 p.m. to complete roughly five ninety-minute sleep cycles you need for an energized next day.

- At 10:00 p.m., dock your phone, turn off the TV (or lower the brightness and make sure you're sitting ten feet from the screen), and shut down laptops.
- Instead of staring at a screen, listen to an audiobook or a podcast. Read a biography or light fiction. If you use an e-reader, put it on the dimmest setting or use blue-light-blocking glasses. Talk to your family (unless they are Energy Vampires). Do paperwork or an off-line non-work project.
- Take an Epsom salts and lavender bath ninety minutes to an hour before lights-out.
- If all your new energy is making you frisky, by all means, enjoy sex (alone or with someone else) during the Power-Down Hour and bask in the post-orgasmic glow of the love hormone oxytocin as it swaddles you gently in a cloud of positive energy. But plan accordingly. Get in bed earlier, so that you are ready for *sleep* at 11:00.

DAILY 5×5 PRIORITY: STRETCH AND BALANCE

This week, along with Bounce and Shake movements, you'll start your practice of wake-up Stretch and pre-bed Balance. Every day, after waking up and a quick trip to the bathroom, you will get on the floor with a mat, or on a carpet or rug, and do a series of five yoga poses that will ignite the pilot light of your day. They say, "I'm here, I'm alive, and I'm ready to roll!" You'll close the day with more yoga, doing poses on one foot at a time to center your mind and improve coordination.

- **Stretch.** At 7:00 a.m., do one minute of Child's Pose, one minute of Cat-Cows, one minute of Sphinx, one minute of Dragonfly, and one minute of Cannonball. These five poses will become your daily morning flow.

- **Shake.** At 11:30 a.m., do one minute each of Neck Looseners, Arm Circles, Leg Swings, Crescent Bends, and Trunk Twists, with a few seconds of rest in between.

- **Bounce.** At 3:00 p.m., do two minutes of Ice Skater Jumps. Rest for one minute. Do two minutes of Burpees.

- **Balance.** At 10:55 p.m., make like a Tree for five minutes, switching feet every thirty seconds. If you lose your balance, no big deal! Just try again.

FASTING ENERGY PRIORITY: EARLY LOADING

People who eat the majority of their calories early in the day have a lower BMI than late eaters—even if they eat the exact same number of calories. So eat a hearty first meal, and gradually decrease the quantity of food as you progress through the day so your final meal is relatively small.

- For **Meal 1** at 10:00 a.m., have bacon and eggs and a green juice or smoothie.

- For a **Snack** at 2:00 p.m., don't overthink it. Have the most practical option at your office or home, like leftovers, a piece of fruit, or a cup of Greek yogurt. If that means having a sandwich, have it, with only the bottom slice of bread. If you crave a candy bar or chocolate chip cookies, have a cup of black coffee to curb your appetite anytime before 3:00 p.m. After that, it will disrupt your sleep.

- For **Meal 2** at 5:30 p.m., have a salad or cooked vegetables with a piece of lean protein. Remember, you *do not* have to eat the entire plate of food. You can save leftovers for your next meal. Smaller plates make for a smaller waist! And once you take that last bite, eat nothing more until breakfast.

JULIA'S WEEK TWO ENERGY SCORE: 4.5 (+1 FROM THE PREVIOUS WEEK)

"I was supposed to do four movement sessions per day, but I averaged two. My alarms went off on schedule, but sometimes I was just too busy and then I lost my chance. It was easy to do Stretch and I got my daughters involved with that, too, which was fun and bonding. My stress level was on the rise by midmorning, and Shake, when I did it, really calmed me down. And Skipping was fun and energizing. It really picked me up. I forgot to Balance a few times, and I felt different getting into bed. So I got out of bed and did Tree so that I could return to bed with a clear head."

Week Three

SLEEP ENERGY PRIORITY: GET OUT OF YOUR HEAD!

When life gets complicated, Slow Bears get anxious and tend to ruminate, or think about problems (sometimes obsessively) they can't necessarily control. Ruminating at bedtime is the leading cause of insomnia for most humans, including Bears. To avoid anxiety, you'll reach for your melatonin-inhibiting devices for distraction, wind up watching TV too late, and then get stuck in a pattern of sleep deprivation. Being crushed under the burden of sleep debt is exhausting, especially for Bears, who really need seven to eight hours per night.

We recommend a two- or three-minute session of diaphragmatic breathing—deep belly breaths—during your Power-Down Hour to help prepare your body for sleep and block rumination before it begins.

↑ Calming Diaphragmatic Breathing for Beginners

1. Sit in a comfortable chair or lie flat on the floor on a mat.
2. Place one hand on your belly just beneath your ribs. Place the other hand on your sternum.
3. Inhale deeply through the nose into your belly until your hand rises for a count of four. Your chest should remain still.

4. Hold for four.
5. Exhale through pursed lips and let your belly and hand deflate for four. Keep exhaling until all the air is expelled.
6. Repeat ten times. Keep it slow and steady.

DAILY 5×5 PRIORITY: BUILD

It's time to Build to increase muscle mass and energize your body during the evening hours when you'd otherwise be doing nothing. You're a Bear, not a Sloth! You'll also practice by-now-familiar Stretch and Balance, as well as Shake and Bounce.

- **Stretch.** You know the drill by now. At 7:00 a.m., do your spine-limbering flow. One minute of Child's Pose, one minute of Cat-Cows, one minute of Sphinx, one minute of Dragonfly, and one minute of Cannonball.
- **Shake.** At 11:30 a.m., increase circulation with your joint-loosening flow. One minute each of Neck Looseners, Arm Circles, Leg Swings, Crescent Bends, and Trunk Twists, with a few seconds of rest in between.
- **Bounce.** Today, at 3:00 p.m., Skip for two minutes, walk for one, and then Skip for another two. If you can't do this outside, do it in place.
- **Build.** At 8:00 p.m., while watching TV, increase muscle mass (and burn more fat) with one minute each of Squats, Crunches, Dips, and Wall Sits, with a few seconds of rest in between.
- **Balance.** At 10:55 p.m., try two minutes of Tippy Toes, two minutes of Figure Four, and one minute of Tree. If you feel at all frustrated, anchor yourself with your hand—and don't forget to breathe!

EMOTIONAL ENERGY PRIORITIES: HOW TO USE MUSIC AND LAUGHTER TO RUN A MARATHON

Slow Bears can use laughter and music as needed whenever they want to improve sleep, mood, and energy. All Bears love to be among people, so if you feel down, reach out to friends and family and make that energizing human connection. Your peak emotional rhythm aligns perfectly with dinnertime, so make plans to eat with friends this week.

We recommend socializing at least three times per week. The number is not random. Stanford University professor and happiness researcher Sonja Lyubomirsky, PhD, told Bustle, "We did a study recently where we asked people to engage in three more social interactions each week for a month than they would normally do. People felt more connected and happier when they did." The interaction could be as casual as a quick lunch with a friend, a quick FaceTime call, a walk together; any social interaction makes people happier. Combining socializing with outdoor fitness or movement multiplies your energizing boost.

If anyone in your social networks is sucking energy and positivity from your life, cut contact with them to instantly feel an emotional lift. If the people on your vampire watch list can't be removed from your life, minimize contact as much as possible.

⚡Energize Tip: Say your friends want to meet for dinner at 7:00 p.m. Should you cancel because that's outside your eating window? No! Go out with friends. Have a great time! And the next day, get back on track. It's okay to break the window once or twice a month if you stick with it most of the time. The emotional boost you get from having fun is greater than the drain of eating late on occasion.

⚡Energize Tip: Steer clear of toxic Emotional Vampires who love to hear about your failures, puncture your good moods, inflict guilt, and sap your energy by complaining, blaming, shaming, and rambling about their problems. Emotional Vampires choose their victims carefully. They go after people who are too nice to tell them to go away, like bighearted Bears. Just say, "Sorry, I don't have the bandwidth for this now."

JULIA'S WEEK THREE ENERGY SCORE: 5 (+.5 FROM THE PREVIOUS WEEK)

"Slow, steady gains. I really missed my nightly popcorn, but only emotionally. I wasn't hungry anymore, but I felt a lack. So I decided to replace one comfort ritual with another and started to take more baths, which is nice. It's better for my body and brain than eating hundreds of calories. If I thought that eating less would make me feel tired, I was completely wrong. The opposite is true."

Week Four

SLEEP ENERGY PRIORITY: THERE'S A NAP FOR THAT

Feeling tired in the afternoon is *normal* and *natural* for Slow Bears. Acknowledging your chronorhythm is energizing in and of itself. Heeding it is the key to your personal power. So, on those drowsy days when Bounce won't be enough to revitalize you after lunch, take a power nap.

For Bears, a short nap (no more than twenty minutes) will restore you to morning levels of energy and alertness and give you a productive afternoon. If you go over twenty minutes, though, you'll come out of a nap groggy and stay that way for an hour.

> ✳ Energize Hack: The best time to nap is approximately seven hours after waking, so if you wake at 7:00 a.m., nap at 2:00 p.m.

Try Michael's Nap A Latte technique (see page 69), or, if you'd rather, just do some diaphragmatic breathing in a darkened room or go outside and do a few extra Bounce exercises instead of napping.

DAILY 5×5 ENERGY PRIORITY: GO HARDER

This week, up the intensity.

- **Experiment.** Are there any movements you have been afraid of trying, like Eagle, Dancer, or Gorilla Burpees? This is the week to face your fears and experiment with the movements that you've avoided.
- **Intensity.** Level up. If you have been coasting with Easy level, go Medium. If you've been doing Medium, go Hard. If you've been doing Hard, up the amplitude. Jump higher, Skip faster, Squat deeper.
- **Duration.** Extend your Daily 5×5 to a Daily 5×7. Add two minutes to each session if you are feeling it!

- **Frequency.** If you've only been doing three sessions a day — some days are just like that — really push this week to hit the magic number five.

FASTING ENERGY PRIORITY: SYSTEMS CHECK

After three weeks of intermittent fasting, you have already adapted to the change in your eating rhythm and have probably noticed that your pants are a bit looser. This week, see if you can push your window by thirty minutes of additional fasting. If you just *can't,* okay. But you won't know unless you try.

- **Open** your eating window at 10:15 a.m.
- **Close** it at 5:45 p.m.

JULIA'S WEEK FOUR ENERGY SCORE: "CAN I SAY 11?" (ACTUALLY, IT'S 7.5, +4 SINCE WEEK ONE)

"It's been a month, but I feel like I have my life back. That emotional shift might be the most significant change in my energy level, but I know I wouldn't feel it if I hadn't made changes in my eating, sleeping, and moving. It all goes together."

She came in with three energy goals:

- **To lose weight.** "I knew my pants were looser, but I didn't want to step on a scale every day because that stressed me out. So I waited until the end of Week Four. I started at 170 pounds, much to my dismay. I finished the month at 159. I lost 11 pounds! I haven't been in the 150s for a decade. So happy, and motivated!"
- **To sleep better.** "I get seven hours per night, and I do get an extra hour on Sunday. The main difference is that when the alarm goes off, I'm up. And at bedtime, I'm able to drift off. It's more than I could have asked for."
- **To move more.** "I'm doing four sessions a day, not five. I've got to be honest about that. But the days I move more, I feel so much better! As a family, we're out and about much more, taking evening walks and kicking

a soccer ball around at the park. In our house, I realize that I set the tone. If I move, so does everyone else. We're transitioning as a family into active, outdoors people. So this has been great for me, and for the people I love."

The weight loss has been an enormous emotional gain for Julia. For every pound dropped, she gained energy and a sense of accomplishment. It can't be overstated that results bring motivation and inspiration. As long as Julia continues to energize her life, she'll stay on a figurative perpetual motion machine and will generate her own power.

Another life energized. Our work here is done.

TROUBLESHOOTING

Slow Bears, a few things to watch out for:

1. **Feeling hemmed in.** Instead of thinking you're a slave to a schedule, understand that real freedom comes from increased energy, dropping extra weight, better communications, and clearer focus. Freedom is about excelling and advancing in your life. If that means keeping track of when you eat, sleep, and work out, it's a small price to pay for limitless possibility. Before long, it'll be automatic.

2. **Slacking off.** It's all too easy to skip a session and say, "I'll hit it twice as hard next time." The point of the five daily sessions is to move consistently throughout the day. So even if you do go hard next time, you'll still miss out on the benefits of moving consistently. Even one minute of movement is better than none.

3. **Peer pressure.** When friends and family complain about your new eating and drinking schedule, it's easy to drift into "just this once." But "just this once" turns into "all the freaking time" in a blink. Instead of letting others pull you away from energizing, bring them into the fold! Then you can have early dinners and afternoon coffee breaks together.

Daily Schedule

The end of Week Four is the beginning of the rest of your high-energy life. Stick with it, and you'll always feel the power.

7:00 a.m. Wake up; 1 minute Child's Pose; 1 minute Cat-Cows; 1 minute Sphinx; 1 minute Dragonfly; 1 minute Cannonball; drink 500 milliliters of water

10:15 a.m. Eating window opens

11:30 a.m. Level up movements — double time optional: 1 minute Neck Looseners; a few seconds rest; 1 minute Arm Circles; a few seconds rest; 1 minute Leg Swings; a few seconds rest; 1 minute Crescent Bends; a few seconds rest; 1 minute Trunk Twists; drink 250 milliliters of water

2:00 p.m. Power nap (optional)

3:00 p.m. 5 minutes Medium or Hard jumps of your choice; drink 250 milliliters of water

5:00 p.m. Exercise (optional)

5:45 p.m. Eating window closes

8:00 p.m. Level up movements — double time optional: 1 minute Squats; a few seconds rest; 1 minute Crunches; a few seconds rest; 1 minute Wall Sits; a few seconds rest; 1 minute Dips; a few seconds rest; 1 minute Kicks; drink 250 milliliters of water

10:00 p.m. Power-Down Hour; turn off all screens

10:55 p.m. 2 minutes Dancer; 2 minutes Eagle; 1 minute Figure Four

11:00 p.m. Bedtime

CHAPTER 9

The Medium Wolf Power Protocol

Vince is a forty-two-year-old financial planner in Boston. His energy story has changed dramatically over the last couple of years. "Even though my mornings were pretty much lost to brain fog, I made up for it in the afternoons and I could stay up and work all night, easily, if I needed to," he said. "My brain doesn't really kick in until the end of the workday, though, which is a problem. My boss thinks I procrastinate when I just can't concentrate fully until after lunch."

He tried to make himself fall asleep faster by smoking cannabis before bed. "I've been a pot smoker since I was sixteen, and I'd never used it for medicinal purposes before," he said. "I wish it made me fall asleep! I get hungry and stupid but weed doesn't really flip the switch between awake and unconscious for me. I'd wake up in a total brain fog, then chug coffee to wake up. I drink *so much* coffee. Probably four or five cups a day, all day long."

Vince continued, "I used to be a runner, but I blew out my knee and had to stop. Running was my stress-relieving outlet, and all of a sudden, it was gone. The pressure at work and the end of exercising was like a one-two punch. All of a sudden, a cloud of exhaustion settled over me and it would not budge. I felt anxious and stressed-out all the time. The youthful energy I relied on—the 'I'll sleep when I'm dead' attitude—dried up. I felt like I had nothing in the tank."

Fortunately, he has a supportive husband who'd read *The Power of When* and diagnosed Vince as being a Wolf, not an insomniac or a wastecase. "I was expected to function on a Bear's schedule, and that had been stressing me out for years. No wonder I was exhausted! Whenever I feel intense pressure, my mood falls off a cliff. My boss's recent criticism of my work had me living with an unshakable uneasy feeling. I probably should have started looking for a new job, but the very idea exhausted me."

Vince's energy goals:

- Reclaim the levels of energy he took for granted when he was younger.
- Figure out how to function as a Wolf on a Bear's schedule.
- Rely less on cannabis and caffeine.

WATCH OUT FOR THE BIGGEST ENERGY DRAINS FOR MEDIUM WOLVES

Many of Vince's behaviors are sucking his energy dry.

↓ Energy Drain: Getting Stoned to Sleep: It's Not About Getting High

As a sleep aid, cannabis does work…*for insomniacs*.[1] Wolves might have trouble falling asleep from time to time, like Bears and Lions. But there's a difference between the medically diagnosable dysfunction of chronic insomnia and being misaligned with your chronorhythm. Vince is not a Dolphin. He was just trying to force his body to fall asleep when he was not tired. Cannabis probably isn't going to knock Medium Wolves unconscious unless they smoke or vape when their melatonin is flowing and their body temperature drops, around midnight.

Although it might cut sleep onset (drift time) by thirty minutes for non-insomniacs,[2] general use of cannabis can be a net energy drain for many. It's dehydrating[3] and reduces saliva production[4] (ahem, cotton mouth, anyone?). We already lose a liter of water just from breathing while asleep. Smoking or vaping cannabis compounds causes dehydration and will have you waking up the next morning as energetic as a dried-out husk.

Vince, and most Medium Wolves, struggles with morning brain fog. Cannabis has been found to lengthen deep, slow-wave delta sleep, the stage for physical restoration. But it shortens REM sleep, the stage when your brain consolidates memory and takes out the cognitive trash.[5] There are two brain receptors that turn on and off when you smoke and/or vape, CB1 and CB2. If you use at night, your CB1 receptors—the groggy ones—don't turn off by morning,[6] and the result is a "pot hangover." For Medium Wolves who are already under pressure to function like Bears in the morning, cannabis is counterproductive. Also, recent research shows that Wolves can have issues with frequent cannabis use.[7]

↑ Energy Gain: Cut Way Back, Dude

For a University of Michigan study, researchers recruited 98 participants in their twenties, assessed their chronotype, and collected data about their cannabis use and sleep quality.[8] The daily weed users' Insomnia Severity Index and sleep-quality scores were nearly double those of nonusers, and about 50 percent higher than those of non-daily users. This was true for morning and evening types, although Wolves are far more likely to use recreational drugs than Lions.

> ✹ Energize Hack: Limit cannabis for weekend use to stop disrupting sleep, improve sleep quality, and make it fun again. Pot hangover cures: sunlight, exercise, and water instead of coffee.

↓ Energy Drain: Living in a Bear's World

Medium Wolves are exhausted because they have to function in a Bear's world. You don't feel tired until 1:00 a.m. and would prefer to sleep until 9:00, but if you have a nine-to-five job—a schedule that's perfect for Bears—you have to wake earlier. As a result, you have to contend with chronic sleep deprivation and the soul-sucking feeling of forcing yourself to be something you're not.

↑ Energy Gain: Getting the Most Out of What You've Got

Chrono-misalignment is a Medium Wolf's number one energy drain. Unless you have workplace flexibility, you have to do what you can to get as much high-quality sleep as you can during a limited number of hours.

To do that, you will have to follow our recommended protocol for creating a Power-Down Hour that starts at 11:30 p.m. every single night. Since Wolves are energetically aroused at that time, the last thing they want to do is turn off all their devices and mindfully slow down their active minds. But if you can make the mental shift and appreciate that being off-screen for an hour before bed is not a fate worse than death, you can use that hour to do so much good for your mind and body. And getting more sleep per week is well worth the sacrifice.

↓ Energy Drain: Shunning Sunlight

Wolves come alive at night. During the day, they tend to hole up in their dens. We know Wolves who barely open their shades before 3:00 p.m. and wear sunglasses all morning because the sunlight hurts their tired eyes. The irony is, sunlight traveling into the eyeballs, along the optic nerve, into your brain's master circadian clock is energizing and can give you the extra boost you need to feel more awake and alert.

> ✴ Wolf Energize Hack: Sunlight gives you energy. No matter how cool you look, never wear sunglasses in the morning.

Sun deprivation not only pushes your chronorhythm further into the dark; it can also cause seasonal affective disorder (SAD), aka "winter blues," an exhausting dip into depression during the cold, dark months. Research has found that, of all the chronotypes, Wolves are more likely to suffer from SAD.[9] In general, Wolves are unfortunately more likely to suffer from mood disorders.

↑ Energy Gain: Don't Be a Vampire

Your Medium Wolf Power Protocol recommendations for sleeping, moving, and eating will address SAD and lift the symptoms of other mood disorders like anxiety, burnout, and nonseasonal depression. Of all types, we know Wolves can have a proclivity for depression. Taking these small steps can ward off exhausting sadness during winter and other seasons of the year:

- Set a consistent wake time.
- Go outside for five minutes in the morning and fifteen minutes in the afternoon (sunglasses *off*) — walk the dog, make a call, listen to a podcast.
- Shut off blue-light-emitting screens an hour before bed.

YOUR NEW KICK ASS ALL DAY, EVERY DAY PLANNER

Program these alarms on your phone for these key times:

8:00 a.m. Wake up; Stretch, plus water

10:00 a.m. Bounce, plus water

10:30 a.m. Eating window opens

4:30 p.m. Shake, plus water

7:00 p.m. Exercise (optional)

8:30 p.m. Eating window closes

9:00 p.m. Build, plus water

11:30 p.m. Power-Down Hour; turn off all screens

12:00 a.m. Balance (no water)

12:30 a.m. Bedtime

Week One

Begin on Monday. Any Monday. THIS Monday.

SLEEP ENERGY PRIORITY: RISE AND LEARN HOW TO SHINE

The objective for setting consistent wake times and bedtimes is to train your brain to follow a sleep rhythm that guarantees enough high-quality sleep and helps you to drift off quickly. The first, most important step to that end is getting up at the same time each day.

- **Wake time is 8:00 a.m.**
- **Do not hit the snooze button.** *Repeat: Do not hit snooze! Force your-self to get out of bed.*
- **Get some sunshine.** Stepping outside or even just sticking your head out the window to get some direct sunlight sends the rise-and-shine message along your optic nerve, straight into your master-clock brain center.

DAILY 5×5 PRIORITY: STRETCH AND BALANCE

Wolves love novelty, and starting a movement schedule is excitingly new. However, you're also a bit rebellious, and not always diligent about stick-ing to regimens. We'll make it super easy for you to stick with yours this week, with daily sessions only at the beginning and end of each day — wake-up Stretch and pre-bed Balance. (See Chapter 4 for the exercises or go to MyEnergyQuiz.com for videos.)

- **Stretch.** At 8:00 a.m., do one minute of Child's Pose, one minute of Cat-Cows, one-minute of Sphinx, one minute of Dragonfly, and one minute of Cannonball.
- **Balance.** In every aspect of energy, Medium Wolves need to *bal-ance* drains and gains. Push yourself but don't burn out. Eat for pleasure but not too much. Party hard but treat yourself gently. Balance will help you find energizing equilibrium in your life. At midnight, do five min-utes of Tree Pose, alternating feet every thirty seconds. And don't forget to breathe!

FASTING ENERGY PRIORITY: ESTABLISH AN INTERMITTENT FASTING RHYTHM

We recommend the eating window of 10:30 a.m. to 8:30 p.m. Only ten hours to eat, no more, and you'll stop eating four hours before bedtime at 12:30 a.m., preventing any gastrointestinal issues from distracting you as you're trying to fall asleep.

VINCE'S WEEK ONE ENERGY SCORE: 5

"I don't find the intermittent fasting schedule to be difficult at all. After a week of morning Stretch, I have made an important discovery. I've been a runner my whole life. I have never tried yoga or done anything to stretch my ligaments and muscles. It's possible my knee problems could have been avoided if I'd had a more balanced approach to fitness. My spine feels loose and amazing. Huge difference in only a week. I am very energized and excited not to limp though my morning with lower back pain. So thanks for that! I still wake up groggy and I know I'm not sleeping long enough, though."

Week Two

SLEEP ENERGY PRIORITY: POWER-DOWN HOUR

When most of the world is already asleep or getting in bed, you are still wide-awake. If you try to fall asleep before you're tired, you'll only get frustrated and anxious. So stay up for a bit, and use that time to quiet the mind before your 12:30 a.m. bedtime.

All Wolves should begin their Power-Down Hour at 11:30 p.m. and allow the deceleration process to happen. The objective is to lower blood pressure and heart rate, a sign of sleep readiness, so you can make a smoother transition from your energetic peak into rest. A drop in core body temperature is another circadian signal that it's time to sleep. It might seem counterintuitive to take a hot bath at this time. But when you step out of the hot water into the colder air, your sudden drop in temperature is a trigger to release melatonin. The ideal room temperature for sleep is 60 to 67 degrees Fahrenheit, so turn on the AC or open a window an hour before bedtime to chill the space.

Lower the temperature of your mind, too, by getting off all devices. Your highly active creative brain needs to disengage well before bedtime, or your thoughts will keep racing. Stop reading stimulating articles and blogs. The melatonin-suppressing blue light from an iPhone two inches from your face doesn't help either. Turn it all off. Don't worry about missing something important. Any news that breaks before morning will still be there when you wake up.

DAILY 5×5 PRIORITY: BOUNCE AND SHAKE

This week add 10:00 a.m. Bounce to clear mental fog and strengthen your body's most energizing muscle: the heart. A strong heart pumps blood and oxygen throughout the body. Better circulation means each cell in the body is nourished and ready to *go* and *do*. By 4:30 p.m., you are at peak mental focus and might not realize you've been working for many hours straight without moving. At that time, Shake things up to get blood circulating to stiff muscles and joints. Weather and season permitting, we strongly recommend that your morning Bounce and afternoon Shake happen *outside*. Sunshine and fresh air are circadian-rhythm reinforcing and mood elevating.

- **Stretch.** At 8:00 a.m., do your flow of Child's Pose, Cat-Cows, Sphinx, Dragonfly, and Cannonball: one minute each.
- **Bounce.** At 10:00 a.m., jumping for five minutes will clear any remaining wisps of brain fog. Do two minutes of Jumping Jacks; one minute of rest; and two minutes of Skipping up and down the block or in place.
- **Shake.** At 4:30 p.m., a circuit of one-minute Neck Looseners, one-minute Arm Circles, one-minute Leg Swings, one-minute Crescent Bends, and one-minute Trunk Twists will work magic on your joints and circulation. A few seconds of rest between movements. Do these outdoors if possible!
- **Balance.** At midnight, do three minutes of Figure Four, alternating feet every thirty seconds. Then do two minutes of Tippy Toes.

FASTING ENERGY PRIORITY: STOP STRESS EATING

Wolves are pleasure seekers and more likely to indulge their hunger than other chronotypes. But if you have noticed any extra weight on your athletic body, intermittent fasting will help take it off. It is more effective than calorie restriction for weight loss.[10] Think of it as a whole new, thrilling way to feel good.

However, Wolves are prone to anxiety and might reach for junk food to soothe their freaked-out feelings of anger, stress, and frustration. It's not your fault. When you're stressed, your body pumps out cortisol, which increases your appetite for high-fat, high-sugar food. Once your insulin goes up in response, hunger hormone ghrelin joins the party, and you're riding a biochemical roller coaster that runs through your pantry over and over again. The bitter irony is that eating all that "comfort" food prolongs cortisol's effect. Internally, you're even more inflamed.

> ⚡ Energize Tip: Stop the mindless shoveling before you take that first potato chip by committing to a Daily 5×5 schedule.

Stress eating—mindlessly shoveling in food to override negative feelings—can be a tough habit to break. Committing to an eating rhythm helps, as does getting up and moving your body.[11] If you keep up with your movement schedule, your cravings for high-fat, high-sugar junk food will diminish.

VINCE'S WEEK TWO ENERGY SCORE: 6.4 (+1.4 FROM THE PREVIOUS WEEK)

"I feel great! So many unexpected benefits of doing this program. My husband is my micromovement coach and we are having a blast doing those crazy yoga poses. Happy Baby? It's a riot. Skipping like an idiot up and down the block? I have to stop from laughing. We're having fun, which is more energizing than anything else. My sleep is better. I can't seem to turn off my devices at 11:30 p.m., though. That's a bridge too far. So I'm compromising by reading my Kindle in the bath with just a few hits of weed."

Caution: Cannabis, electronics, and water do not mix!! Vince was counseled to buy a print book and use his cannabis after he got out of the tub (and he ended up using less because he was already soooo relaxed).

Week Three

SLEEP ENERGY PRIORITY: BEDTIME BREATHWORK

So far, you've worked on waking up consistently and establishing Power-Down Hour practices. This week, focus on bedtime itself—getting in bed at a consistent time each night, and doing some breathing exercises to prevent rumination and shorten drift time.

Drift time is how long it takes you to fall asleep after you get in bed. Ideally, it'll be no more than twenty minutes. Taking a bath, lowering the room temperature, and turning off screens all help to prepare you for the drift.

Try a deep belly-breathing technique to get you over the consciousness hump. This kind of breathwork improves focus, raises positive affect, and lowers cortisol levels.[12] It also stimulates the vagus nerve, which runs from your neck down to your belly. The vagus nerve turns on the switch for the rest-and-digest parasympathetic nervous system and turns off the fight-or-flight sympathetic nervous system. Since your fight-or-flight hormones go up at night, this is the perfect soothing exercise for you.

↑ **Calming Diaphragmatic Breathing for Beginners**
1. Sit in a comfortable chair or lie flat on the floor on a mat.
2. Place one hand on your belly just beneath your ribs. Place the other hand on your sternum.
3. Inhale deeply through the nose into your belly until your hand rises for a count of four. Your chest should remain still.
4. Hold for four.
5. Exhale through pursed lips and let your belly and hand deflate for four. Keep exhaling until all the air is expelled.
6. Repeat ten times. Keep it slow and steady.

DAILY 5×5 PRIORITY: BUILD

This week, add Build movements. As a Medium Wolf, you can increase muscle mass easily, but you have to put in the effort to do it. By working your biceps, triceps, quads, and abs with Build, you'll increase your energy exponentially. It's the final push to get you over any exhaustion hurdle.

- **Stretch.** At 8:00 a.m., do the flow you now know by heart: Child's Pose, Cat-Cows, Sphinx, Dragonfly, and Cannonball: one minute each.
- **Bounce.** Jolt yourself fully awake at 10:00 a.m. with a challenging two minutes of Burpees, one minute of rest, and two minutes of Ice Skater Jumps.
- **Shake.** Ideally, you'll Shake for five minutes after every hour of sitting. At the very least, at 4:30 p.m., stand up and run through a five-minute circuit of Neck Looseners, Arm Circles, Leg Swings, Crescent Bends, and Trunk Twists. One minute each with a few seconds of rest in between.
- **Build.** More muscles mean faster fat burn. So, at 9:00 p.m., do a minute each of Squats, Crunches, Dips, Wall Sits, and Kicks with a few seconds of rest in between.
- **Balance.** At midnight, experiment for two minutes each with Eagle and Dancer, alternating feet every thirty seconds. Then seal your day with a triumphant minute of Tree Pose.

EMOTIONAL ENERGY PRIORITY: MAKE THE MIND-BODY CONNECTION

Medium Wolves seem to live in extremes. When you focus, you go deep into the flow zone. When you socialize, you are the light and laughter in any room. Your emotions are big and raw, and when you put pressure on yourself, it can feel crushing. Michael is a Medium Wolf and he's experienced the phenomenon of believing he could push himself so far, for so long, only to have his body give out. If he hadn't had his cardiac event, he would have burned out and had to come back from a different kind of physical trauma and devastating emotional symptoms.

Your emotional goal this week is to de-stress. Everything in the Power Protocol is designed to flood your body with healthy, happy hormones and reduce a tsunami of cortisol and adrenaline. But along with fasting, being conscientious about sleep hygiene, and *moving,* add a daily dose of laughter, music, time with loved ones, or Netflix (even though Wolves are extroverts, they recharge from alone time as well).

> ✴ Energize Insta-hack: Change your Energy Posture by sitting up straight to boost your mood and mindset.

Your feelings are a mirror of your posture. If you are slumped over, your energy will shrink as well. But if your body is held upright and takes up space, your mood and energy will correspondingly expand.

Don't believe us? It's been scientifically proven. For a study of 90 men and 55 women, researchers from San Francisco State University asked half the participants to slouch and half to sit upright while they recalled stressful memories.[13] That one factor—how they sat—determined how they recovered from mental tension. As you can probably guess, the slouchers got stuck in their negative thoughts. The backs-straight, heads-held-high group significantly reduced the duration and severity of their negative thoughts. Just. Like. *That.*

Medium Wolves, you can get stuck in your own head. To get out of stressful brain loops, do a Bounce session (the best time to add one is at 7:00 p.m.) or simply sit up straight. You'll feel better, and that alone is energizing.

VINCE'S WEEK THREE ENERGY SCORE: 7.3 (+.9 FROM THE PREVIOUS WEEK)

"I had a sleep breakthrough this week. I got in bed at midnight, was asleep within a half an hour, and woke up with a clear head. I can't say for sure what change has turned the tide. Maybe it's because I'm getting back into fitness. The five daily sessions aren't like taking an after-work run, which used to be my jam. But the short sessions of movement seem to be adding up. I have noticed that the slight belly pouch I put on since I stopped running has gone down. All told, it's been a good week. And my newfound energy is, apparently, noticeable. People at work have asked me what's gotten into me lately. I tell them, 'I'm feeling energized—you should check out this book I'm reading!'"

Week Four

DAILY 5×5 ENERGY PRIORITY: GO HARDER

This week, up the intensity.

- **Experiment.** Are there any movements you have been afraid of trying, like Eagle, Dancer, or Gorilla Burpees? This is the week to face your fears and experiment with the movements that you've avoided.
- **Intensity.** Level up. If you have been coasting with Easy level, go Medium. If you've been doing Medium, go Hard. If you've been doing Hard, up the amplitude. Jump higher, Skip faster, Squat deeper.
- **Duration.** Extend your Daily 5×5 to a Daily 5×7. Add two minutes to each session if you are feeling it!
- **Frequency.** If you've only been doing three sessions a day — some days are just like that — really push this week to hit the magic number five.

FASTING ENERGY PRIORITY: SYSTEMS CHECK

After three weeks of intermittent fasting, you have already adapted to the change in your eating rhythm and have probably noticed that your clothes are a bit looser. This week, push your window by fifteen minutes on either end to test your endurance. If you just *can't*, okay. But you won't know unless you try.

- **Open** your eating window at 10:45 a.m.
- **Close** it at 8:15 p.m.

VINCE'S WEEK FOUR ENERGY SCORE: 8.2 (+3.2 SINCE WEEK ONE)

Vince started this program with some energy goals:

- **Reclaim the levels of energy he took for granted when he was younger.** "I feel great, period. That's all I can say. I'm channeling my energy into this process, and it's working. The best part by far is how my energy is contagious. My colleagues seem more 'up' around me. My marriage is definitely enjoying the benefits. I feel like I bring happiness and power with me wherever I go, and people respond to it. In this state of body and mind, I'm going to apply for some new work opportunities and see what happens. And if I get rejected, all I have to do to turn around dark thoughts is to sit up straight. Magic!"

- **Figure out how to function as a Wolf on a Bear's schedule.** "I'm getting enough sleep and use movement to clear my head in the morning so I can be more productive. I'm not at my best until afternoon, but now I can make use of my mornings, and my boss has noticed. I'm definitely doing better as a Wolf in a Bear's world."

- **Rely less on cannabis and caffeine.** "I just had to cut way back on both, and I found that when I was using less of one, I needed less of the other. I was using pot to help me sleep and coffee to wake me up. But neither one actually did what I hoped it would do. Now, I get high on the weekends for a long walk with my husband, and we giggle like idiots the whole time. And I drink coffee to get me over energy dips in the late morning, early afternoon. It's all good!"

Such a joyful success story for Vince, a man who noticed a problem and took steps to fix it. Having agency—feeling empowered to change things for the better—is super energizing, and it's available to anyone who has the confidence to take it. Starting Week One of your Power Protocol puts you on the path toward finding the confidence and energy to change your life. Just take that first baby step in good faith, and you're already there.

Another life energized. Our work here is done.

TROUBLESHOOTING

Medium Wolves, a few things to watch out for:

1. **Rebelliousness.** Wolves do not like to be told what to do. We're giving you a lot of rules to follow, and your natural disposition is

to question authority. Make the subtle mental adjustment to see your Power Protocol as helpful information instead of rigid instruction. That way, you've got nothing to rebel against.

2. **Frustration.** Some of the movements will feel awkward until you practice them enough. Ice Skater Jumps are hard to master. But who cares if you feel like a goof? Perfection is not the point. Silence the Olympic judges in your mind and give yourself top marks for effort.

3. **Backsliding.** After a few good days or weeks, you might think, "I've earned a break." Exhausting yourself with overeating and being sedentary is no reward; it's a self-inflicted punishment. And then one "off" day rolls into another, and you're right back where you started. Cheat days only cheat you. Fight backsliding by meeting it with an energizing action. Do something that will lift you up (sex, or dancing, or cooking a fantastic meal with friends).

Daily Schedule

The end of Week Four is the beginning of the rest of your high-energy life. Stick with it, and you'll always feel the power.

8:00 a.m. Wake up; 1 minute Child's Pose; 1 minute Cat-Cows; 1 minute Sphinx; 1 minute Dragonfly; 1 minute Cannonball; drink 500 milliliters of water

10:00 a.m. Level up movements — double time optional: 5 minutes jumps of your choice

10:45 a.m. Eating window opens

4:30 p.m. Level up movements — double time optional: 1 minute Neck Looseners; a few seconds rest; 1 minute Arm Circles; a few seconds rest; 1 minute Leg Swings; a few seconds rest; 1 minute Crescent Bends; a few seconds rest; 1 minute Trunk Twists; drink 250 milliliters of water

7:00 p.m. Exercise (optional)

8:15 p.m. Eating window closes

9:00 p.m. Level up movements — double time optional: 1 minute Squats; a few seconds rest; 1 minute Crunches; a few seconds rest; 1 minute Wall Sits; a few seconds rest; 1 minute Dips; a few seconds rest; 1 minute Kicks; drink 250 milliliters of water

11:30 p.m. Power-Down Hour; turn off all screens

12:00 a.m. Level up — double time optional: 5 minutes Balance poses of your choice

12:30 a.m. Bedtime

CHAPTER 10

The Slow Wolf Power Protocol

Ann, a fifty-five-year-old writer in Brooklyn, came to Michael with a long list of problems. "I'm prediabetic with an A1c of 5.9," she said. "I have acid reflux that tortures me at night and is one of the reasons I lie awake until 1:00 or 2:00 a.m. Another reason I can't just drift off is I'm always worrying. It's become a bad brain habit. The light goes out and my anxiety turns on, and it just snowballs the longer I lie in bed awake. My doctor said I need to lose at least 20 pounds to help bring down my blood sugar. But I have zero time to exercise because I've been the sole earner for my family since my husband retired, and I've got one kid in college and another in grad school."

Clearly, Ann was under a great deal of pressure and stress—and that's a dangerous situation for a Wolf. Of all the chronotypes, Wolves are the most likely to suffer from mood disorders like depression.[1] "I've had bouts of depression in the past, but nothing too scary," she said. "I'm more concerned about my body than my mind, although I get it that they're connected. My brain hasn't let me down the way my body has. When I got that diagnosis of prediabetes, it came as an absolute shock. I have to get that under control!"

Like a typical Wolf, Ann feels foggy in the morning and comes into her power in the afternoon. "I'm all but useless before noon. By the time I really get in the groove, around 3:00, the day is half gone and the pressure to produce goes way up. I basically chain myself to my desk and don't leave until I've done enough," she said. "It's a lot of sitting, and a lot of coffee drinking to maintain. I work ten or twelve hours a day. I tried to use a standing desk when my spinal stenosis got really bad—did I mention my chronic neck pain?—but I found it exhausting. Along with coffee, I snack often to keep myself going. I always reward my day of work with a night-time snack of Cheerios with oat milk. I might top off the bowl once or twice. But it's not like I'm parked in front of the TV for hours drinking beer and shoveling brownies."

Ann's energy goals:

* Lose weight.
* Lower blood sugar.
* Stop acid reflux.
* Get more done.
* Feel less stressed-out.
* Fall asleep faster.
* Move *at all*.

Slow Wolves, you might start this process in the worst energy state of all the Power Profiles. Slow Wolves are the most challenged in every energy area. Along with being *less* motivated to move, and living *more* sedentary lives, Slow Wolves are probably carrying an exhausting 10 to 20 extra pounds. That said, of all the Power Profiles, Slow Wolves have the most to gain. And we have found that they make the most dramatic improvement over a month. So turn on your inner cheerleader voice, and *go, go, go!*

WATCH OUT FOR THE BIGGEST ENERGY DRAINS FOR SLOW WOLVES

Many of Ann's behaviors are sucking her energy dry.

↓ Energy Drain: Eating Late

Remember how your parents told you, "Nothing good ever happened after midnight, so be home by then"?

Creatures of the night are hungriest after dark. Wolves' hormonal circadian rhythm makes them crave carbs at 9:00 p.m., 10:00 p.m., or later. Ann's late-night snack of plain Cheerios with oat milk is only 150 calories, but it's got 25 grams of sugar. She admitted to having two or three bowls, racking up the calories and sugar grams to the point where her "good" snack might as well be a beer and a brownie. (Her daytime snack of fruit isn't helping her either. Fruit has fiber, but it's also packed with fructose. For a prediabetic, an apple is not much better than a cookie.)

Slow types are keenly sensitive to carbs and will turn excess glycogen into fat. Ann wasn't burning those carbs since they were consumed at the end of the day. Extra weight and a heavy carb diet are risk factors for diabetes. If she crossed the line from prediabetic to type 2 diabetic—indicated by an A1c of 6.5; Ann is only 0.6 percentage point away—she'd have an increased risk of heart disease and certain cancers along with diabetes symptoms and risk factors.

Even though your chronorhythm makes you hungry later in the day, eating at night also exacerbates acid reflux. Stopping eating four hours before bed allows the body to "rest and digest" and reduce acid reflux symptoms. Late eating can disrupt or delay sleep, too. Sleep deprivation causes every physical and mental health condition that Ann is worried about, and increases her cravings for *more carbs*.

↑ Energy Gain: It's Called a Food Curfew

Slow Wolves, if you do nothing else but break the late-eating habit, your energy score will go way up. We recommend an eight-hour eating window followed by sixteen hours of fasting. Since you're not hungry in the morning anyway, start eating at noon, and stop at 8:00 p.m.

Late snacking can turn into binging—one bowl of cereal becomes three. Binging is the main reason Slow Wolves tend to be above average weight.[2] Drawing a line on all post-dinner snacking is the only solution. There is no "a little nibble" for Wolves, who have the lowest impulse control of all chronotypes. Know it, accept it about yourself, and work with that knowledge. We've worked with Slow Wolves who, on this eating/fasting schedule, lost 20 pounds in two months, improved their insulin sensitivity, reversed prediabetes, and felt like new (younger, happier) people. You can do it, and you'll be so glad you did.

↓ Energy Trap: Sitting, the New Smoking of Bad Habits

The sedentary life causes chronic exhaustion, carb cravings, weight gain, diseases, sleep disruption, and decreased cognitive clarity. A parked ass really does trigger a mental and physical feedback loop from hell. If your brain is tired (from sleep deprivation), movement will feel like a painful chore. It's human nature that if you hate doing something, you're less likely to do it. And this is never more true than for rebellious-by-nature Wolves, who don't see why they should do anything they're unenthusiastic about.

↑ Energy Gain: Get Up Off of That Thing and Move

In the Slow Wolf's Daily 5×5 flow, you'll Stretch first thing, which signals to the body and mind, "I am awake and moving now"; then Bounce in the midmorning, which acts like a shot of espresso. Shake your way into your energetic peak in the afternoon. With your abundant evening energy, you'll be motivated to Build on the day's effort and gain strength. Finally, you'll incorporate Balance into your Power-Down Hour, to quiet your active mind so that you can sleep.

Since the movement sessions are spread out, you won't be sedentary for longer than a few hours at a time. And since you only have to do them for a few minutes each, you don't have to carve out a huge chunk of time from your already-compressed day. Plus, you'll do two of these sessions before you're fully awake. Sleepwalk through them, as long as you go through the motions. Over time, you may actually look forward to them.

↓ Energy Trap: Getting in Bed Too Early

Ann's thinking makes sense, that if she fell asleep earlier, maybe she'd have more energy. But it doesn't work that way for Wolves. When you get in bed on a Bear's schedule — at 11:00 p.m. — you'll wind up lying awake, with increasing anxiety about why you're not falling asleep. Anxiety feeds on itself, and you'll fill up with nervous energy (aka "autonomic arousal") about what will surely be an exhausting day tomorrow. Your anxiety will turn into a self-fulfilling prophecy. Chrono-misalignment — living out of sync with your chronorhythm — causes anxiety-based insomnia for Wolves.

↑ Energy Gain: Just Do You!

Live according to your own circadian rhythm and get in bed at 12:30 a.m. You'll drift off faster and the anxiety of lying awake for hours will be a nonissue.

MICHAEL SAYS . . .

I don't get in bed until 12:00 a.m. because I know what will happen if I try to "make" myself fall asleep when I'm not tired. I will wind up sleeping fewer hours than I would have if I'd just waited until my body signaled to me that I was ready for rest. Remember, sleep is a lot like love—the more you look for it, the less likely it is to show up! How can you tell you're genuinely tired and ready for sleep?

Yawning every few minutes.

Eyelids feel heavy.

Longing to lie down.

I've had patients who've tortured themselves for years, believing they were severe insomniacs, when it turned out they were extreme Wolves. They didn't have a sleep problem; they were just not following their natural-born chronorhythm. When they shifted their bedtime, the nightly battle to fall asleep ended. They nodded off quickly, instead of ruminating for hours about not sleeping.

They still had to wake up on a Bear's schedule, but their net sleep gain was increased by seven or eight hours per week. *They gained an entire night's worth of sleep by getting in bed an hour LATER.* It's counterintuitive, but it works.

YOUR NEW KICK ASS ALL DAY, EVERY DAY PLANNER

Program these alarms on your phone for these key times:

8:00 a.m. Wake up; Stretch, plus water

11:30 a.m. Bounce, plus water

12:00 p.m. Eating window opens

4:30 p.m. Shake, plus water

7:00 p.m. Exercise (optional)

8:00 p.m. Eating window closes

9:00 p.m. Build, plus water

11:30 p.m. Power-Down Hour; turn off all screens

12:00 a.m. Balance (no water)

12:30 a.m. Bedtime

Week One

Begin on Monday. Any Monday. THIS Monday.

SLEEP ENERGY PRIORITY: RISE AND LEARN HOW TO SHINE

The objective for setting consistent wake times and bedtimes is to train your brain to follow a sleep rhythm that guarantees enough high-quality sleep and helps you to drift off quickly. The first, most important step to that end is getting up at the same time each day.

- **Wake time is 8:00 a.m.**
- **Do not hit the snooze button.** *Repeat: Do not hit snooze! Force yourself out of bed.*
- **Get some sunshine.** Stepping outside or even just sticking your head out the window to get some direct sunlight sends the rise-and-shine message along your optic nerve, straight into your master-clock brain center.

DAILY 5×5 PRIORITY: BUILD

A common goal for Slow Wolves is to lose weight. The motivation might be to fit into smaller pants—who doesn't want that?—but the real benefit of carrying a lighter load is having more energy. The problem for Slow types is that your body is primed to hold on to fat for dear life. Being sedentary means there's no opportunity for change. To turn that trend around and get results ASAP, focus this week on increasing muscle mass. Muscles burn through dietary glycogen incredibly fast, so increasing muscle mass will have you shift into fat-burning mode more quickly (not as fast as Fast or Medium types, but faster for *you*).

Tonight, at 9:00 p.m. when you're feeling your most energetic, Build those muscles and the weight will drop off. (See Chapter 4 for the exercises or go to MyEnergyQuiz.com for videos.)

> ⚡ Slow Wolf Energize Tip: If you feel the gravitational pull of inactivity—and it will be strong—counter it with a new mantra. As you inhale, say, "More muscles," and as you exhale, say, "More energy"; or be like the Rock and just tell yourself, "Earn it, every fucking bit."

Build. At 9:00 p.m., do Squats, Crunches, Wall Sits, Dips, and Kicks. One minute each with a few seconds rest in between. If possible, do them again at 10:00 p.m. But if you can only manage one five-minute set, that's great. One minute of movement means one less minute of being sedentary. Moment by moment, your self-confidence will soar!

FASTING ENERGY PRIORITY: ESTABLISH AN INTERMITTENT FASTING RHYTHM

We recommend the eight-hour eating window of 12:00 p.m. to 8:00 p.m. Only eight hours to eat, no more. For healthy insulin response and to prevent diabetes, it's essential that you give your digestive system a break by fasting for sixteen hours straight.

We're not going to lie. This week, you will probably feel hungry at night. You're used to munching between 8:00 p.m. and midnight. To achieve your energy goals, that has to stop. By day four, it'll be much easier.

An energize buddy can help you get through the rough start. And if you do slip, don't beat yourself up. Just try again. With every failed attempt comes an opportunity to *succeed*!

ANN'S WEEK ONE ENERGY SCORE: 3.1

"Well, I know that my emotional energy was way down this week because I was so 'hangry' at night! I hoped I'd feel an energy boost from the Build movements—I did enjoy those and they did distract me from my hunger—but when I couldn't have my cereal, I just felt sad, like I'd lost a beloved friend. In happier news, my acid reflux wasn't as bad. In sleep news, my husband pointed out that my two or three nightly trips to the bathroom to pee had dropped to just one. I don't think I got more hours, but the quality went up. This might actually work, a little."

Week Two

SLEEP ENERGY PRIORITY: POWER-DOWN HOUR

At 11:30 p.m., when most of the world is already asleep or getting in bed, you are still wide-awake. If you tried to fall asleep now, you'd only get frustrated and anxious. We recommend a 12:30 a.m. bedtime for you.

Your brain is still on fire with bright ideas during the final hour of the day. If you give yourself the slightest cognitive encouragement, you'll stay alert and engaged for one or two hours more. We recommend that all Wolves begin their Power-Down Hour at 11:30 p.m. and allow the "mental deceleration process" to happen. The objective is to lower blood pressure, a sign of sleep readiness, so that you can make a smoother transition from your energetic peak into rest.

One science-backed activity for your Power-Down Hour is to take a steaming bath. According to a new Japanese study of more than 1,000 people, soaking in a hot tub every night can control blood pressure, lower BMI, and lower A1c,[3] the Holy Grail for diabetics. An hour before bed, slip into the tub with calming Epsom salts and lavender. If you don't want to take a bath every night, shift your morning shower to nighttime.

A drop in core body temperature is another circadian signal that it's time to sleep. It might seem counterintuitive to take a hot bath to do this. But when you get out of the tub or shower into the colder air, the sudden change will trigger a melatonin release. According to the National Sleep Foundation, the ideal room temperature for sleep is 60 to 67 degrees Fahrenheit, so turn on the AC or open a window an hour before bedtime to chill the space.

Lower the temperature of your mind, too, by getting off all devices.

Your highly active, creative brain needs to disengage well before bed-time, or your thoughts will keep racing. Stop reading stimulating arti-cles and blogs. The melatonin-suppressing blue light from an iPhone two inches from your face doesn't help either. Turn it all off. Don't worry. Any news that breaks before morning will still be there when you wake up.

DAILY 5×5 PRIORITY: BOUNCE AND BALANCE

This week, add Bounce to strengthen another muscle: the heart. A strong heart pumps blood and oxygen throughout the body. Better circulation means that each cell in the body is nourished and energized. You'll also incorporate a key Power-Down Hour brain-quieting ritual: Balance.

- **Bounce.** At 11:30 a.m., while you're still half-asleep, jumping for five minutes will clear any remaining wisps of brain fog. You'll feel more awake and might even get a few things done before noon. Keep it simple: two minutes of Jumping Jacks, one minute rest, and two minutes of Jumps for Joy.
- **Build.** At 9:00 p.m., do your nightly muscle strengtheners: one minute each of Squats, Crunches, Dips, Wall Sits, and Kicks, with a few seconds rest in between. Do it again at 10:00 p.m. if motivated to do so.
- **Balance.** At midnight, do five minutes of Tree Pose. Alternate feet every thirty seconds. If you lose your balance, no big deal. Just get back into the pose. And don't forget to breathe.

FASTING ENERGY PRIORITY: COMFORT VS. CARBS

Intermittent fasting gets easier the longer you do it. A month from now, the very idea of snacking after dinner will make you queasy. Last week might've been a shock to your system. As a night creature, with your brain churning and hungry for nutrients long after dark, you crave carbs. Exhausted people indulge those cravings, at the expense of sleep, weight, and energy. But your brain will do just fine late at night if it burns stored fat for energy instead of dietary carbs. So when those cravings hit — and they might come and go for another couple of weeks — answer them with emotional comfort instead of comfort food.

What will make you feel comfy and cozy besides munchies after dark?

Sex?

Build moves?

Hot tea?

A Netflix movie? *Property Brothers? Downton Abbey?*

FaceTime with your kids or friends?

Make a list and keep adding to it. Run through it to counter cravings at night.

MICHAEL SAYS...

This was the toughest part for me. I always felt like a failure once I started snacking, and then I usually just said, "Fuck it, I'm worth it," and would proceed to eat, well, everything. Having the cutoff time was very helpful for me, but I did several other things to help ensure my success. You should try them, too:

1. Have healthy, portion-controlled snacks — a handful of raw almonds, an apple, popcorn — set up each week that, if you do eat, won't have you ringing up 1,200 calories and a ton of sugar.
2. Turn off all the lights in the kitchen. It is a "pattern interrupt." When you turn off the lights, it stops the mindless part of mindless eating. Then see if you can...
3. Sip water. Many people have no idea if they are hungry or thirsty, so sip some water to figure it out. (Not too much before bed, though.) If you are using cannabis, remember it will make you both hungry *and* thirsty. (Maybe hold off until the weekend?)

4. Enlist an accountability partner. Explain to someone you love, who has a slightly watchful eye, what you are trying to do. They may be very helpful! I had one couple get really energized when one wanted to break his fast and his partner offered sex instead. Guess what happened?

5. Eat when you should. You have a feeding window, so use it! It works, we promise!

ANN'S WEEK TWO ENERGY SCORE: 5 (+1.9 FROM THE PREVIOUS WEEK)

"The fasting went much better this week. Not eating is starting to feel normal and not like a horrible deprivation. I've been enjoying creating a nightly ritual around peace and disengagement and have consciously shifted away from obsessing about work toward free-form thinking, like daydreaming at midnight. My mind is still working, but it's not spinning. I'm still lying awake until 1:30 a.m., but my stomach and my brain are quieter. I feel a bit thinner, and definitely more hopeful."

Week Three

DAILY 5×5 PRIORITY: STRETCH AND SHAKE

This week, start the daily practice of wake-up Stretch to limber yourself up for an energized day, and afternoon Shake to counteract the circulatory effects of sitting for too long.

- **Stretch.** Upon waking at 8:00 a.m., after a quick trip to the bathroom, do some spine work. Get to know this five-minute flow: one minute each of Child's Pose, Cat-Cows, Sphinx, Dragonfly, and Cannonball.
- **Bounce.** Jolt yourself with energy at 11:30 a.m. with two minutes of Burpees, one minute of rest, and two minutes of Ice Skater Jumps.
- **Shake.** Ideally, you'll Shake for five minutes after every hour of sitting. At the very least, at 4:30 p.m., stand up and run through a circuit

of one minute each of Neck Looseners, Arm Circles, Leg Swings, Crescent Bends, and Trunk Twists to increase circulation to your muscles and joints. Take a few seconds rest between moves.

- **Build.** More muscles mean faster fat burn. So does your routine of one minute each of Squats, Crunches, Dips, Wall Sits, and Kicks starting at 9:00 p.m. Try again at 10:00 p.m. Every minute of Building adds up.

- **Balance.** At midnight, experiment with two minutes of Figure Four and two minutes of Tippy Toes. Then seal your day with a triumphant one-minute Tree Pose.

EMOTIONAL ENERGY PRIORITY: MOOD BOOSTER

Listening to music, laughing, giving, and socializing will all keep your mood up this week. As a Wolf, you already know the energizing power of enjoying all that life has to offer. And as a pleasure seeker and risk taker, you don't need to be reminded that fun is energizing. A disco playlist at 11:30 a.m. will energize your Bouncing and boost your mood while you do it. A comedy podcast during your Power-Down Hour will lower blood pressure and help prepare your body for sleep.

But even a Will Ferrell movie marathon can't undo the dampening effect of inactivity. Moving reduces stress and elevates mood on a biochemical level. Being sedentary makes you feel bad and sad, and energy follows your moods all the way to the ground. Emotional exhaustion rolls into anxiety, which causes sleep deprivation, flowing into physical tiredness and all-round general misery. The best mood tonic, and energy infusion, for Slow Wolves is getting off the chair and onto your feet.

According to a study of 202 adults with major depression by researchers at Duke University, exercise was as effective as antidepressant medication for relieving symptoms.[4] This is not to say that people with mood disorders should stop taking their prescribed medications. (Never stop taking medication without the approval and supervision of your doctor.) As a supplemental therapy, movement lifts you up ... along with heading off mood and energy killers like sleep deprivation and overeating.

For every other Power Profile, more energy starts with sleep. Sleep is crucial for you, too. But your golden ticket for energy gains begins with moving every chance you get.

ANN'S WEEK THREE ENERGY SCORE: 6.1 (+1.1 FROM THE PREVIOUS WEEK)

"Feeling good! I have the eating thing down, and this week, I really focused on moving more. Granted, I'm not running loops around the park. But just standing up and shaking out my limbs in the afternoon has been a revelation. I honestly did not think it would have any effect; *wow*, was I wrong. I used to power through hours of work and it always felt like a slog. But after a Shake session, when I sit back down, it's like starting over fresh each time. I'm not forcing myself to continue. I'm doing what I always did, minus the struggle. I really wish I'd known this in college."

Week Four

DAILY 5×5 ENERGY PRIORITY: GO HARDER

This week, up the intensity.

- **Experiment.** Check the movement list in Chapter 4. Are there any you have been afraid of trying, like Dancer Pose or Gorilla Burpees? This is the week to face your fears and experiment with the movements that you've avoided.
- **Intensity.** Level up. If you have been coasting with Easy level, go Medium. If you've been doing Medium, do Hard. If you've been doing Hard level, up your amplitude. Jump higher, Skip faster, Squat deeper.
- **Duration.** Extend your Daily 5×5 to a Daily 6×5, or a Daily 5×7. Add a session or minutes if you are inspired.
- **Frequency.** If you've only been doing three sessions a day—some days are just like that—really push this week to hit the magic number of five.

FASTING ENERGY PRIORITY: SYSTEMS CHECK

After three weeks of intermittent fasting, you have already adapted to the change in your eating rhythm and have probably noticed that your clothes are a bit looser. This week push your window by fifteen minutes on either end to test your endurance. If you just *can't,* okay. But you won't know unless you try.

- **Open** your eating window at 12:15 p.m.
- **Close** it at 7:45 p.m.

ANN'S WEEK FOUR ENERGY SCORE: 8 (+4.9 SINCE WEEK ONE)

Ann started this program with a bunch of energy goals:

- **Lose weight.** "Intermittent fasting is magic," she said. "I've been doing it for four weeks, and only suffered for about four days. But I stepped on the scale and I've lost 8 pounds! I will never go back to mindless snacking again. You don't have to count calories or even carbs. Just count hours. It's so much easier this way. I'm making progress after feeling powerless to change for a long, long time."
- **Lower blood sugar.** "It's too soon to tell. I need to wait until I've been doing this for three months to get an accurate A1c reading. But losing weight was what my doctor said I had to do. I've done it and will continue to. I have also seen my sweet tooth relax a bit, so I'm crossing my fingers!"
- **Stop acid reflux.** "Staying hydrated and not eating for four hours before bed: acid reflux magic."
- **Get more done.** "Well, I got a lot done in my pre-Energize life, but now I don't feel so exhausted doing it. It's a huge difference to feel better all day long, instead of feeling like my entire life is a slog."
- **Feel less stressed-out.** "Movement is the cure for stress. I get it now. I have the tools I need. So whenever I feel low, I just have to stand up and jump or skip."

- **Fall asleep faster.** "My sleep is definitely better, and not getting in bed before I'm yawning every minute or two has helped a lot with drift time."

- **Move *at all.*** "I'm moving throughout the day. But in a large sense, I'm moving forward in my life. I felt stuck for a long time. But now I'm in control of my life again. I can change. I have that power. I can move myself, and mountains."

Another life energized. Our work here is done.

TROUBLESHOOTING

Slow Wolves, a few things to watch out for:

1. **Impatience.** Slow types have the most to gain from this program, but it'll take commitment. Your metabolism is like a cruise ship; you can spin the wheel, but it takes some time to change direction. We guarantee that you will lose weight and feel supercharged. Give it two weeks.

2. **Frustration.** Some of the movements will feel awkward until you practice them enough. Ice Skater Jumps are hard to master. But who cares if you feel like a goof? Perfection is not the point. Silence the Olympic judges in your mind and give yourself top marks for effort.

3. **Rebelliousness.** Wolves do not like to be told what to do. We're giving you a lot of rules to follow, and your natural disposition is to question authority. Make the subtle mental adjustment to see your Power Protocol as helpful information instead of rigid instruction. That way, you've got nothing to rebel against.

Daily Schedule

The end of Week Four is the beginning of the rest of your high-energy life. Stick with it, and you'll always feel the power.

8:00 a.m. Wake up; 1 minute Child's Pose; 1 minute Cat-Cows; 1 minute Sphinx; 1 minute Dragonfly; 1 minute Cannonball; drink 500 milliliters of water

11:30 a.m. Level up movements—double time optional: 5 minutes jumps of your choice

12:15 p.m. Eating window opens

4:30 p.m. Level up movements—double time optional: 1 minute Neck Looseners; a few seconds rest; 1 minute Arm Circles; a few seconds rest; 1 minute Leg Swings; a few seconds rest; 1 minute Crescent Bends; a few seconds rest; 1 minute Trunk Twists; drink 250 milliliters of water

7:00 p.m. Exercise (optional)

7:45 p.m. Eating window closes

9:00 p.m. Level up movements—double time optional: 1 minute Squats; a few seconds rest; 1 minute Crunches; a few seconds rest; 1 minute Wall Sits; a few seconds rest; 1 minute Dips; a few seconds rest; 1 minute Kicks; drink 250 milliliters of water

11:30 p.m. Power-Down Hour; turn off all screens

12:00 a.m. Level up—double time optional: 5 minutes Balance poses of your choice

12:30 a.m. Bedtime

The Fast and Medium Lion Power Protocol

Leslie is a forty-five-year-old executive at a large, national nonprofit charitable organization, and is, frankly, "addicted to work and working out," she said. "I wake up at 5:00 a.m. every day, whether I want to or not. It's not like I flutter my eyes open and have to psych myself to get out of bed. It's more like my eyes snap open and I'm *up*. There is no way I could just lie there and wait for the rest of the world to wake up. My day begins when it's still dark outside. I'm divorced now, but when I was married, my husband could sleep for two or three more hours. By the time he was awake, I'd already have gone for a run, showered, had breakfast, and done an hour's worth of work."

She takes pride in her discipline. "I'm very fit and ambitious," said Leslie. "My ex is a little pudgy and too content about where he is professionally. I might've made one too many jokes about his laziness, but we were fundamentally incompatible in a lot of ways, including our different sleep schedules. If I don't exercise, I go crazy. He hated the gym and only plays sports with friends. I liked hanging out, just the two of us, and he always pushed us to make plans with other people. I know now that he is a Slow Bear and I'm a Fast Lion. Opposites attracted...until we grew apart."

Leslie's day starts like a rocket, shooting straight up to the sky, and then gradually descends until the 8:00 or 9:00 p.m. crash. "I can get so much done during the day, and I feel incredibly lucky to have those two or three hours of total productivity in the morning when everyone else is still asleep. Sometimes I laugh, picturing a colleague waking up and seeing my flurry of 6:00 a.m. emails," she said. "But I miss out on so much when my energy hits a wall at 9:00 p.m. or earlier. It feels like I live in my own world. I can't binge-watch TV shows like everyone else and rarely go to

> **Fast and Medium Lions,** of all the Power Profiles, come into our program with a higher energy baseline than any other type.

events and concerts unless they start sharply at 7:00 p.m. I have to beg off parties and late dinners with friends. Since the divorce, I feel more isolated than ever. I'm an introvert and need my alone time. But this is ridiculous."

Fast and Medium Lions tend to lead a healthy lifestyle already, exercising and eating well. Like Leslie, they might overdo it with exercise and too-careful eating. Leslie said, "I can be obsessive about my diet and I read tons of wellness books. I read *The Power of When* by Dr. Breus, and it was a big relief to understand myself better. But some friends and family have told me that I can be too intense about my eating and fitness. I feel guilty when I fall off whatever program I'm on. If I miss a workout because I'm too tired, I make myself run twice as long the next day. It's possible I'm overtraining, but I don't like to think about it that way."

Leslie knows she's hit the genetic jackpot to have been born a Fast Lion. An early riser, she gets all the positive traits that go with it, especially supreme conscientiousness.[1] With a fast metabolism, she can eat pretty much whatever she wants, even though she chooses to stick to a healthy diet. But all that said, she does have energy goals:

- Prolong her alert hours until 10:00 p.m. so she can socialize and date.
- Be more flexible about her rigid fitness standards, which cause stress and tire her out.
- Tend to her emotional energy, which always comes last.

WATCH OUT FOR THE BIGGEST ENERGY DRAINS FOR FAST AND MEDIUM LIONS

Many of Leslie's behaviors are sucking her energy dry.

⬇ Energy Drain: Just Because You Can Doesn't Mean You Should

Fast and Medium Lions push themselves, hard. They have never met a limit they didn't want to jump over. The feeling of accomplishment is mega-energizing. It's one of the finest, purest sources of personal power we've got. And when you get a hit of that powerful dopamine rush, you want to feel it again and again.

From the outside, overexercisers and perfect-carb-ratio dieters might inspire awe. But exercising too much and being too strict about your food choices drains energy from your body and spirit. Overtaxing muscles, undereating, and under-resting set you up for injury, burnout, chronic stress, and energy depletion. When you push yourself *too* hard, the law of diminishing returns applies.

MICHAEL SAYS...

Overtraining syndrome (OTS) is real. It'll drain your energy, crush your appetite for food and sex, and wreck your ability to think and sleep.[2] When people exercise too much, it causes stress and inflammation, which lead to insomnia. Symptoms include fatigue, poor concentration, greater injury risk, immune system suppression, sleep disruption, dwindling bone density, and heart strain.

An energize keyword is *moderation*. If you love to exercise, by all means, go for it. But then give your body a day to rest and recharge so you can do it again on your strongest footing. Unless you are a professional athlete or training for an ultramarathon, you put your body and brain at risk by doing too much exercise without recovery. Although the habit might be entrenched that you "need" to run every single day, try to wrap your mind around a new kind of energized life.

We know you're up for a challenge, Fast and Medium Lions. You have a lot to do that requires concentration. But if your insides are bathed in cortisol, it's extremely difficult to focus. So for the next four weeks, try a different way, and decide for yourself based on your weekly energy scores how increasing moderate activity and decreasing intense exercise affects you.

Chronically exceeding the Centers for Disease Control's recommendation of two and a half hours of intense exercise per week might cause an injury that takes longer than it should to heal.[3] Overstressed muscles trigger a cortisol release, and that brings about poor-quality sleep, chronic inflammation, and weight gain around the middle.

↑ Energy Gain: Less Is More

Stacey is a Fast Lion, and throughout most of her life, whenever she was presented with a new physical challenge, she accepted it without hesitation. While in recovery from her addiction to drugs, overexercise became her new addiction, but it slowly proved to be harmful, too. Once she turned fifty and started to work on her overall health, sleep, and energy, she realized that less is more when it comes to just about everything, and definitely intense exercise. She had clients who believed that the more they did, the better off they'd be. They showed up for a 6:30 spin class and stayed for the 7:30 class, and only then would they allow themselves to have a cup of low-fat yogurt.

Intense exercise without a day or two of rest between workouts forces your body to work double time to recover and repair broken-down muscles. It's a net energy loss. If you worked out every other day, or every third day, and did moderate activity (like the five daily movements), you'd come out with massive energy gains.

"Less is more" works for dieting as well. No question, Lions tend to eat healthfully.[4] But being too strict with yourself can turn into an obsession, which can develop into an addiction. And any behavior that provokes intense feelings of guilt or shame triggers a cortisol flood to go with your fight-or-flight anxiety response. Cortisol is a Lion's friend first thing in the morning. It wakes you and gives you that delicious raring-to-go feeling that only *your* chronotype enjoys. You don't need another dose of it at the sight of a tempting treat. Instead of fatiguing your adrenal gland with stress, just have the occasional cookie!

↓ Energy Drain: Feeling Like a Freak on Planet Bear

Lions are a breed apart. They rise hours earlier than Bears and Wolves. They have a higher energy baseline. They're performance-driven and conscientious in ways that make other types marvel. In *The Power of When,* Michael wrote about the phenomenon of "Lion Envy." What others might not realize is that, for introvert Lions, it can be very lonely to spend those dark wee hours waiting for the world to wake up. They might be envious themselves of other socially gifted types who come alive at the parties and gatherings they're too tired to attend.

Besides living on a different schedule from most of the planet, they are lacking in one key characteristic that makes their differences even harder on them. According to a study by a leading chronobiology researcher at the University of Warsaw,[5] morning-oriented types rank lower in emotional intelligence — in skills like perception, understanding, assimilation, and managing. So when Lions feel isolated and disconnected, part of that exhausting dynamic might relate to their naturally impaired human-relations deficiency. It's not your fault that things get a bit awkward sometimes. But it is upsetting, and therefore draining.

↑ Energy Gain: It's a Lion's World and You Rule It

The only way to block the emotional drain of feeling like you're a bit out of step with the other chronotypes is to accept it. You're different. So what? Being different makes you special and that's a badge of honor. Despite their lingering "otherness," Lions can and do rise above those feelings. That same expert from Poland did another study of 349 participants that found morning-oriented people show significantly greater "life satisfaction" than their more emotionally intelligent evening-oriented peers, regardless of age or gender.[6]

Lions, you still win. Despite the pressure you put on yourself to achieve, despite feeling somewhat lonely at the top, you are optimistic about your life and feel damn good about it. Next time you recoil from the sting of disconnection, take a deep belly breath, do a session of Bouncing, and remind yourself that you are still king of the jungle, and that all you need is within your reach.

↓ Energy Drain: Even You Can't Run on Empty

Leslie, newly single, would very much like to get out and mingle. But most of the men she meets online or in real life ask her out for late dinner dates when she might do a face plant into her low-cal dessert. She'd like to be able to stay relatively energized until 10:00 p.m. to widen the pool of her potential love interests. But by that time, Lions' body temperature has been on a steep decline for hours, and their melatonin is pumping.

↑ Energy Gain: How Do You Feel About Nooners (Napping or Sex)?

You can't change your DNA, but you can nudge your chronorhythm a bit to squeeze out an extra hour or two of energy in the evening without losing sleep. The hack: midday napping. A half-hour nap at 1:00 p.m. will recharge your body battery enough to get you over the 9:00 p.m. exhaustion wall, according to a recent French study.[7] Not only that; it will boost immunity and reduce stress—something that you go-hard-or-die-trying types need to pay attention to.

> **Energize Hack:** Even if you've had a bad night or are under stress, a daytime nap hits the reset button on your endocrine system and puts you back at the starting line.

And after your nap? Another kind of afternoon delight. Lions' testosterone levels are high all morning.[8] Plus, your "hedonic tone"—your all-day pleasure and good-vibes peak—crests at midday.[9] A sex break in the middle of the day—with or without a nap—would certainly put an extra bit of pep in your step.

YOUR NEW KICK ASS ALL DAY, EVERY DAY PLANNER

Program these alarms on your phone for these key times:

6:00 a.m. Wake up; Stretch followed by 500 milliliters of water

6:15 a.m. Exercise (optional)

7:00 a.m. Eating window opens

9:00 a.m. Shake, plus water

1:00 p.m. Power nap (optional)

2:00 p.m. Build, plus water

6:00 p.m. Bounce, plus water

7:00 p.m. Eating window closes

9:00 p.m. Power-Down Hour; turn off all screens

9:30 p.m. Balance (no water)

10:00 p.m. Bedtime

Week One

Begin on Monday. Any Monday. THIS Monday.

SLEEP ENERGY PRIORITY: THERE'S A NAP FOR THAT

Think of it as a fun game. Make it competitive, since you love to win. This week, at 1:00 p.m., can you adjust your schedule so that you can go into a room, turn off the lights, and lie down for twenty minutes? Just twenty minutes? If you're not in the habit of afternoon napping, it might seem strange to take this midday break. But by allowing yourself to slow down for as little as twenty minutes, and doze or meditate, you are banking battery power that you will feel and appreciate later on.

> Nap Hack: Wake time + seven hours = ideal nap time. For Lions, a twenty-minute nap at 1:00 p.m. will recharge your body battery, boost immunity, lower stress, and buy you an extra hour of energy at the end of the day.

214

DAILY 5×5 PRIORITY: REDUCE DAILY WORKOUTS TO THREE TIMES PER WEEK

Unlike every other type, Fast and Medium Lions will need to slow things down in order to amp up their energy. This week, we're asking you to increase energy by allowing your body to rest and recover for one day after every intense workout. So, if you go for a run on Monday, skip it on Tuesday. Plan your week in advance. We recommend a Monday, Wednesday, Saturday schedule. Add each workout to your calendar. During rest days, use the time you would have been exercising for moderate or low-impact physical activity, like walking, vacuuming, or dancing around your living room. And remember, you'll still be doing five movement sessions on the days you don't exercise.

Of the Daily 5×5 sessions, this week, you'll open your day with Stretch and seal the day with Balance. Lions will feel a sense of accomplishment if they bookend their day with positive behaviors. (See Chapter 4 for the exercises or go to MyEnergyQuiz.com for videos.)

- **Stretch.** After waking up at 6:00 a.m. and a quick trip to the bathroom, get on the floor next to your bed and do Child's Pose for one minute, Cat-Cows for one minute, Sphinx for one minute, Dragonfly for one minute, and finish up with one minute of Cannonball. By the end of the week, you'll be able to do this progression on autopilot.
- **Exercise.** This week, plan to do intense exercise on Monday, Wednesday, and Saturday only, at 6:15 a.m.
- **Balance.** As a new part of your pre-bed winding-down ritual, at 9:30 p.m. do Tree Pose. Stand on one foot for thirty seconds, then switch sides. Keep alternating feet until five minutes have passed. If you lose your balance, just try again. And remember to breathe!

FASTING ENERGY PRIORITY: ESTABLISH AN INTERMITTENT FASTING RHYTHM

We recommend the twelve-hour eating window of 7:00 a.m. to 7:00 p.m. Only twelve hours to eat, no more. Although weight control and the looming threat of type 2 diabetes are probably not your problems, your intermittent fasting plan ensures circadian consistency. Regular routines promote quality sleep so when you wake up like a shot at 6:00 a.m., you'll feel rested and ready to tackle your day.

LESLIE'S WEEK ONE ENERGY SCORE: 6

"Compared to my sister, who is a Bear, I know I'm coming into this program with a much higher energy starting point than most people. But that doesn't mean I don't want *more* energy! I tried napping and succeeded on workout days. I guess I do need to recharge when I push myself. It was very difficult not to go running every day. I didn't know what to do with myself all morning, so I wound up lacing up and doing my same loop in the park, but just walking. When running, I listen to hip-hop and stay focused on my speed. When walking, I listened to an audiobook and looked at the birds and people. I was covering the same ground, but the feeling and experience were completely different. The aha! moment was about the importance of mixing things up. I always want to do more, and walking showed me that 'more' isn't necessarily faster or harder. 'More' can be slower and softer. Less stress and self-inflicted pressure is definitely energizing. If I keep having these moments, my whole life is going to get better! Yes, I am always annoyingly positive when I start something."

Week Two

SLEEP ENERGY PRIORITY: POWER-DOWN HOUR

At 9:00 every night, shut down all screens. Unplug your mind from stressful thoughts or anything that might cause a surge of anxiety. Read a book, listen to music or a podcast, do some easy chores or tasks around the house, talk to a friend, take a bath, have sex! Turn down the temperature of your mind to prepare your body for sleep. Balance poses, accompanied by mindful breathing exercises, will help you get calm and quiet.

SLEEP HACK: GUAVA LEAF TEA

Along with being antioxidant-rich, guava leaf tea has been found to boost immunity, lower cholesterol, prevent and treat diabetes[10]—and improve sleep quality. One reason you might wake up during the night, Fast and Medium Lions, is low blood sugar. That sets off a hormonal cascade that winds up turning on the cortisol tap. Cortisol is one of the circadian hormones that signal your master clock to start a new day, so it's not ideal for it to flow at 3:00 a.m. Guava leaf tea keeps blood sugar stable, so that hormonal cascade never starts, and you stay asleep or calm down enough to fall back to sleep. We recommend the brand GuavaDNA, available online.

DAILY 5×5 PRIORITY: BOUNCE AND BUILD

Along with continuing your reduced workout schedule and doing morning Stretch and pre-bed Balance, this week, add afternoon Build to take advantage of your still-high energy, and evening Bounce to force your body into a second wind when your energy starts to drag.

- **Stretch.** At 6:00 a.m., run through your five-minute progression: Child's Pose for one minute, Cat-Cows for one minute, Sphinx for one minute, Dragonfly for one minute, and finish up with one minute of Cannonball.
- **Build.** At 2:00 p.m., you are already feeling pumped, so channel that power into muscle development. Do one minute of Squats, with a few seconds of rest. Then do one minute of Crunches, with a few seconds of rest. Finish with one minute of Dips.
- **Bounce.** Your energy trend line is on the decline at 6:00 p.m. Flatten it with one minute of Jumping Jacks followed by twenty seconds of rest. Then do one minute of Jumps for Joy, and a short rest. Then two minutes of Skipping (in place or around the house).
- **Balance.** At 9:30 p.m., during your Power-Down Hour, root into your bedroom floor by doing Tree Pose, alternating feet for three minutes. Then try Tippy Toes for two minutes.

FASTING ENERGY PRIORITY: DON'T FORGET TO EAT DINNER

A recent scientific review of some thirty-six studies on dietary habits of morning-oriented types (Lions) vs. evening-oriented types (Wolves) found that Lions are more likely to have a lower BMI, eat a healthier diet, have a lower risk for eating disorders, eat most of their calories earlier in the day, and consume less alcohol. Lions with a fast or medium metabolism can pretty much eat whatever they want and maintain a healthy weight, in part because their gut hormones (like "Feed me!" ghrelin and "I'm full" leptin) are as speedy as they are. Wolves' and Bears' same hormones are delayed, meaning they don't get hungry until later in the day, and there's a longer time gap between eating and feeling satisfied.[11] Once again, you win the genetic lottery.

But just because your eating habits are naturally favorable doesn't mean they can't be improved upon for more energy. Fast and Medium Lions can be intense about their work and responsibilities, and if they're in the zone, *they might forget to eat in the afternoon and evening*. To sustain your energy, you need to consume calories throughout the day, or your tank will run dry. Also, when the stomach is empty — three or four hours after eating your last bite — it tells your brain that it's time to feel sleepy. By having that last bite at late-for-you 7:00 p.m., you can delay your sleep rhythm by an hour or more.

LESLIE'S WEEK TWO ENERGY SCORE: 6.6 (+.6 FROM THE PREVIOUS WEEK)

"So many changes, and things to do. On paper, it seems like a lot to keep track of. But by programming my phone to alert me whenever I'm supposed to do something, it's easy to manage. My biggest hurdle this week: eating so late. I'm used to eating at 5:00 p.m. and getting in bed at 9:00 p.m. Now I'm trying to have dinner at 7:00 p.m. and stay up until 10:00 p.m. Obviously, the Power Protocol is a much friendlier schedule for having a life. But it's an adjustment. I'm just not hungry that late. So instead of having a whole meal at 7:00 p.m., I have a bowl of cereal with berries or, if I'm out, I get an appetizer. It's enough food to trick my stomach into thinking it's full. I guess it's because of changing hormones, but it feels odd now to get in bed an hour after eating. I'm still tired, but I don't want to lie down. The daily movement sessions are excellent. Psychologically, I feel great about doing things all day long, like I'm making up for my scaled-back workout schedule. Logically, I know five minutes of Jumping Jacks are not the same as a three-mile run. But I feel good, and I'm happy to do them."

Week Three

DAILY 5×5 PRIORITY: SHAKE

The final piece — Shake — is added this week. At 9:00 a.m., after you've been hard at work for a couple of hours, stand up from your desk, Shake yourself up, and boost circulation of blood and oxygen to all the joints that have been locked into a seated posture for too long.

- **Stretch.** At 6:00 a.m., do your spine-limbering flow, which you know by heart now: Child's Pose for one minute, Cat-Cows for one minute, Sphinx for one minute, Dragonfly for one minute, and finish up with one minute of Cannonball.
- **Exercise.** Stick with a 6:15 a.m. Monday-Wednesday-Saturday schedule.
- **Shake.** At 9:00 a.m., start with one minute of slow Neck Looseners (easy does it). Do thirty seconds of Arm Circles on each side. Then rest and move on to thirty seconds of Leg Swings on each side. One minute of Crescent Bends. Finish up with one minute of Trunk Twists.
- **Build.** At 2:00 p.m., do one minute each of Crunches, Dips, Squats, and Wall Sits. If you're feeling your power, add one minute of Kicks.
- **Bounce.** At 6:00 p.m., challenge your coordination and heart by doing two minutes of Burpees, followed by a one-minute rest. Then do two minutes of Ice Skater Jumps.

- **Balance.** Advance your Tree Pose at 9:30 p.m. for five minutes, alternating feet every forty-five seconds.

FASTING ENERGY PRIORITY: FOOD CHOICE

Now that your fasting rhythm is established, take a closer look at your food choices. Review the nutrition guidelines in Chapter 5 and be honest about what you're putting in your body.

- **Hydration.** Start the day by drinking at least 500 milliliters of water first thing. After each movement session, have another 500 milliliters.
- **Limit grains and dairy.** Switch to nut or oat milks. Limit rice, pasta, bread, and grains to a total of 200 grams per day.
- **Add more vegetables!** The sky's the limit with leafy greens. Limit starchy veggies like potatoes, squash, yams, and sweet potatoes to one serving per day.
- **Avoid high-sugar fruits** like grapes, melon, citrus, and stone fruit. Switch to high-fiber apples and low-glycemic berries.

EMOTIONAL ENERGY PRIORITY: BREAK A RULE

In a study of 360 Spanish university students, researchers set out to determine if there were significant differences in the ways chronotypes *think* and *behave*, and they uncovered some fascinating data.[12] It turns out that being a morning-oriented Lion correlates with a low fear of pain; being internally focused, realistic, unimaginative (sorry), logical, not particularly innovative, asocial, not outgoing, confident, conventional, dutiful, and controlling; and having a positive general outlook. You gather your thoughts through tangible experience and use your well-honed logic and analytical skills to draw conclusions. Any new info that comes in has to be squared with knowledge you've already accumulated (a way of thinking called "conservation-seeking"). Behaviorally, you are the embodiment of self-control. You respect authority and are cooperative, even though you're formal and aloof socially. The number one personality trait is conscientiousness, so when you set out to do something, you stick with it. We are thrilled you've committed to our Power Protocol and have made huge strides already.

But chasing perfection is a pointless pursuit and a waste of emotional energy. Nothing in life is perfect, and neither are you. So this week, just to relieve any building pressure to be perfect, we're asking you to dutifully, confidently, conscientiously break one "rule." Eat outside of your eating window. Go to bed too early. And then do what you do best: Be realistic about how that purposeful slipup affected your energy. Use the emotional power of your logic and self-control to assess and progress. This will reaffirm your belief in the Energize program and, with that knowledge, you can feel even better about your commitment to it.

> ✷ Energize Hack: Purposefully slip up to see how it impacts your energy. Enter the data into your Weekly Energy Diary and check against previous weeks to see if there is a change in how the slipup makes you feel physically and energetically.

LESLIE'S WEEK THREE ENERGY SCORE: 7.6 (+1 FROM THE PREVIOUS WEEK)

"I've got my sleep routine, eating window, and movement schedule down. I do spend a lot of emotional and mental energy reminding myself about the science behind the plan and how it all fits into my idea about what works for me. I seemed to plateau energetically this week, and it made me question the effectiveness of the plan...until I did the emotional exercise of a forced slipup. I like logic puzzles, like 'By doing X, does Y or Z happen?' I applied it to the program by going running three days in a row. By Wednesday morning, after my third run in a row, I was noticeably more tired and less able to focus. It is hard for me to make big changes in my life, but this experiment showed me that one old view really is wrong. For most of my life, I thought exercising to exhaustion was energizing. As it turns out, doing anything to exhaustion is just exhausting. Not to get into too much detail, but I can see some evidence of this in my marriage as well. This was a big shift in my thinking, and potentially, in my behavior in future relationships."

Week Four

SLEEP ENERGY PRIORITY: GO AHEAD AND WATCH A MOVIE ALL THE WAY THROUGH!

How do you feel at 9:00 p.m.? Are you dragging or able to keep going for another full hour? If your goal was to be functional until 10:00 p.m., look at how far you've come. Did you gain fifteen minutes of alertness in the last few weeks? Thirty minutes?

This week, push yourself to stay awake until 10:00 p.m. using all the techniques we've discussed. Napping. Eating on a fixed 7:00 a.m. to 7:00 p.m. schedule. Allowing screen exposure until 9:00 p.m. That should get you to 10:00 p.m. Seven and a half hours of uninterrupted sleep later, you'll wake up fully charged and restored.

DAILY 5×5 ENERGY PRIORITY: GO HARDER

This week, up the intensity.

- **Experiment.** Check the movement list in Chapter 4. Are there any you have avoided trying, like Dancer Pose or Gorilla Burpees? This is the week to experiment with the movements that you've haven't tried yet.
- **Intensity.** Level up. If you have been coasting with Easy level, go Medium. If you've been doing Medium, do Hard. If you've been doing Hard level, up your amplitude. Jump higher, Skip faster, Squat deeper.
- **Duration.** Extend your Daily 5×5 to a Daily 6×5, or a Daily 5×7. Add a session or minutes if you are inspired.
- **Frequency.** If you've only been doing three sessions a day—some days are just like that—really push this week to hit the magic number of five.

FASTING ENERGY PRIORITY: WATER CHECK

After three weeks of intermittent fasting, your eating might not have changed as drastically as the other chronotypes'. But eating on a regular schedule does wonderful things for your digestion. This week focus on whether you're drinking enough. Start tracking your glasses of water this week to make sure you're well hydrated.

Drinking tons of water is exhausting! Shoot for draining a one-liter bottle twice a day by taking little sips (except at mealtime).

LESLIE'S WEEK FOUR ENERGY SCORE: 8.6 (+2.6 FROM WEEK ONE)

When she started her Power Protocol, Leslie had several goals:

- **Prolong her alert hours until 10:00 p.m. so she can socialize and date.** "I'm able to stay up until 10:00 p.m. Mission accomplished! So if I start dating, I'll be able to say 'yes' to a dinner date and might not fall asleep in the middle of the movie."
- **Be more flexible about her rigid fitness standards, which cause stress and tire her out.** "I've scaled back my grueling runs and have added walking and relaxed bike rides. I can't not move a lot; it's just who I am. But I'm smelling the flowers, as it were, not hell-bent on mowing them all down now."
- **Tend to her emotional energy, which always comes last.** "I have more energy at night to hang out at happy hour with colleagues. The subtle shifts have opened my schedule, and myself, to connect, and that gives me emotional energy as well."

Another life energized. Our work here is done.

TROUBLESHOOTING

Fast and Medium Lions, a few things to watch out for:

1. **A bumpy adjustment.** Lions are at the mercy of their efficiency. It's not easy for them to change their routines or try new things. If you expect to feel a bit out of sorts as you adjust, you'll make a smoother transition, and the extra energy will be worth it.

2. **Going too hard.** We know you love to go hard, but remember that overdoing it defeats the purpose. If your insides are steeped in cortisol, your body will waste tons of energy to deal with the stress. Rest between workouts! Take power naps!

3. **Overlooking energy sources.** You might get so caught up in the physical energy gains, you forget about the emotional ones. We're helping you find more hours in the day so you can spend them with friends and family to make powerful (and empowering) connections. Take advantage of emotional energy opportunities.

Daily Schedule

The end of Week Four is the beginning of the rest of your high-energy life. Stick with it, and you'll always feel the power.

6:00 a.m. Wake up. Hard level of each movement. 1 minute Child's Pose; 1 minute Cat-Cows; 1 minute Sphinx; 1 minute Dragonfly; 1 minute Cannonball; drink 500 milliliters of water

6:15 a.m. Exercise 45 minutes to 1 hour Monday, Wednesday, Saturday

7:00 a.m. Eating window opens

9:00 a.m. Hard level of each movement—double time optional: 1 minute Neck Looseners (easy does it); a few seconds rest; 1 minute Arm Circles; a few seconds rest; 1 minute Leg Swings; a few seconds rest; 1 minute Crescent Bends; a few seconds rest; 1 minute Trunk Twists; drink 250 milliliters of water

1:00 p.m. Power nap (optional)

2:00 p.m. Hard level of each movement—double time optional: 1 minute Crunches; a few seconds rest; 1 minute Squats; a few seconds rest; 1 minute Dips; a few seconds rest; 1 minute Wall Sits; a few seconds rest; 1 minute Kicks; drink 250 milliliters of water

6:00 p.m. Hard level of each movement—double time optional: 2 minutes Burpees; 1 minute rest; 2 minutes Ice Skater Jumps; drink 250 milliliters of water

7:00 p.m. Eating window closes

9:00 p.m. Power-Down Hour; turn off all screens

9:30 p.m. Hard level: 5 minutes Balance poses of your choice

10:00 p.m. Bedtime

CHAPTER 12

The Slow Lion Power Protocol

Jason is the fifty-year-old CEO of a tech firm in San Francisco, and he is on the phone, taking calls, answering texts, writing emails, from dawn (when he wakes up) until he hits his energy wall around 9:00 p.m. Although he spends most of his daylight hours at his desk, in his mind, he's extremely active. "I have a million things going on at any one time. My mind is always moving at light speed. I accomplish more before breakfast than a lot of my staff does all day long. But when I stand up after sitting for hours and my hips ache, I'm always a bit surprised. I've done so much, but I haven't moved more than ten feet from my desk."

He doesn't even leave his desk for food. His assistant keeps meals and snacks coming all day long. "I don't even have to ask. There's always something to eat within reach. I'll admit, I do a lot of mindless shoveling. I know that I need to eat, so I do, but I don't really care that much what it is or how much I consume. And yes, I could stand to lose 15 pounds around the middle. My doctor told me that my life is too stressful and that she's worried about my heart."

As for sleep, Jason does fairly well getting seven hours per night. "In theory, I'd like to sleep a bit more, but when is that going to happen? I'm wide-awake at dawn and I don't get home from work until it's nearly time to fall into bed and pass out."

His energy problem is that since he turned fifty, he's had a steady decline in his baseline energy level. "I've always been aware of my day-to-day energy rhythm, that it starts high and steadily declines until bedtime," he said. "Now my overall health is on a steady, irreversible decline, and it's scaring the crap out of me. Is depleted energy what I have to look forward to? If I can do something to slow down or reverse the inevitable, I'd like to try."

Jason's energy goals:

- To level up his baseline energy.
- To change his movement and eating to reduce body aches and his belt size.
- To lower his blood pressure, blood sugar, and cholesterol.

WATCH OUT FOR THE BIGGEST ENERGY DRAINS FOR SLOW LIONS

Many of Jason's behaviors are sucking his energy dry.

↓ Energy Drain: Optimism Has Its Drawbacks

Jason put off lifestyle changes and self-care because he believed (until now) that he had plenty of time to make necessary adjustments. This optimistic view of time is classically Lionish. According to a recent University of Warsaw study of 316 participants, each chronotype had a different "temporal perspective," aka concept of time.[1] Evening-oriented participants (Wolves) had a pessimistic, even hostile, concept of time. But the morning-oriented ones (you, Lion) were optimistic about it.

Your optimistic view of time *can work against you* when it comes to your health and energy. You believe you have plenty of time to tend to your energetic goals, but . . . you might not. Don't put off taking care of your health.

↑ Energy Gain: How Soon Is Now?

From your perspective, time is on your side. But the reality is, there is no time like the present to take care of your health and boost your energy. For a Spanish study on chronotype and procrastination, 509 adults were assessed for their morning or evening preference, then tested on two aspects of procrastination: indecision and work avoidance.[2] The Wolves tended to be avoidant. They put off things they didn't want to do. The

Lions didn't avoid work, but they did struggle with decision-making. Once you make a decision, you *get right on it*. So we're asking you to put your energy and health at the top of your priorities list for the next four weeks. Once you lock into a commitment, you are all in. Nothing can stop you. You are a force of nature.

↓ Energy Drain: Sitting Is the New Smoking

Here's the thing: Your brain might be racing at a million miles a minute while you are at your desk, controlling the universe. But your body is just sitting there, lump-like. What Slow Lions seem to overlook or forget is that feeling wide-awake and being physically active are not the same thing. Your work life might very well be intense and demanding. You might rightfully believe that you can't walk away from your desk for even a few minutes. But understand that sitting for ten hours or more per day drains energy — and increases your risk of premature death.

↑ Energy Gain: Eleven Minutes Can Save Your Life

A Norwegian study crunched the fitness tracker data from nine previous studies on some 50,000 middle-aged and older people in Europe and the United States.[3] The researchers divided the participants into three groups based on their lifestyle and physical activity level. The top tier moved the most; the middle tier was somewhat active; the bottom third was the most sedentary. They dug into the death registries of people in the studies from a decade earlier. The results confirmed what we already intuitively know: The top tier (who moved the most) had the lowest rate of premature death. The bottom tier (who were sedentary) had a 260 percent increased likelihood of dying early compared to the most active group.

The relevant finding in this study that beautifully confirms what we have been recommending to you: The middle tier, the moderate movers,

> ✦ Energize Hack: Eleven minutes of movement might be the difference between an energized life and early death. By following your Power Protocol, you'll get twenty-five minutes per day, more than twice what you need, without ever breaking a sweat. Nice trade!

logged just eleven minutes of walking on average per day and were *significantly less likely to die before their time* compared to the sedentary bottom tier.

↓ Energy Drain: Eating in the Fast Lane

Lions are always in a rush. But one speedy habit might be dangerous for Lions with a slow metabolism: eating fast. Jason said that he is often impatient about sitting down for meals and rushes through them so he can get back to work. But eating too fast is one of the reasons he, and other slow metabolic types who have a higher risk of obesity and diabetes, is overweight.

A Japanese study of 59,717 type 2 diabetics over a five-year period set out to draw a line primarily between obesity and eating speed.[4] The data showed that the speed eaters had the highest BMI and were the least healthy overall. The moderate-speed eaters were 29 percent less likely to be obese compared to the gobblers; the slow eaters were 42 percent less likely to be obese. What's more, snacking at night and eating within three hours of bedtime also correlated with a sluggish metabolism and high BMI.

↑ Energy Gain: Slow Down and Chew!

As a Lion, you are lucky. Your chronotype DNA puts you at a lower risk for obesity than any other type.[5] However, even Lions need to be mindful about their eating rhythm and food choices. Since you have a slow metabolism, you're naturally prone to having a higher risk of heart disease and diabetes. If you do nothing to change, your already sluggish metabolism will continue to slow down as you get older. But if you take steps to speed things up—by locking into a consistent sleeping, eating, and moving schedule that's right for your chronotype—you stave off some risk of disease and can get your weight under control.

We're not asking you to count how many times you chew every single bite of food. No one has time for that. But try to slow your eating speed. Your last bite of the day is at 6:00 p.m., so set a timer for 5:45 p.m. to start eating, and do not put that last forkful in your mouth until 6:00 p.m.

Force yourself to savor the meal for a full fifteen minutes, and you just might lower your risk of metabolic disorders like diabetes.

YOUR NEW KICK ASS ALL DAY, EVERY DAY PLANNER

Program these alarms on your phone for these key times:

6:00 a.m. Wake up; Stretch, plus water

6:15 a.m. Exercise (optional)

9:00 a.m. Shake, plus water

10:00 a.m. Eating window opens

1:00 p.m. Power nap (optional)

2:00 p.m. Build, plus water

6:00 p.m. Eating window closes

7:00 p.m. Bounce, plus water

9:00 p.m. Power-Down Hour; turn off all screens

9:30 p.m. Balance (no water)

10:00 p.m. Bedtime

Week One

Begin on Monday. Any Monday. THIS Monday.

SLEEP ENERGY PRIORITY: THERE'S A NAP FOR THAT

This week, at 1:00 p.m., can you make an adjustment to your schedule so that you can go into a room, turn off the lights, and lie down for twenty minutes? If you're not in the habit of afternoon napping, it might seem strange to take this midday break. But by allowing yourself to slow down for as little as twenty minutes, and doze or meditate, you are banking battery power that you will feel and appreciate later on.

> ✺ Nap Hack: Wake time + seven hours = ideal nap time. For Lions, a twenty-minute nap at 1:00 p.m. will recharge your body battery, boost immunity and lower stress, and buy you another hour of energy at the end of the day.

↑ Fast Refresh Meditation for Beginners

If you can't nap, try meditation to get a comparable reset. Here's a starter guide:

1. Sit on a pillow or comfy chair, or lie on the bed or on a yoga mat on the floor.
2. Close your eyes; use an eye mask if you need it.
3. Breathe naturally. Don't try to count or control at first.
4. Focus on the breath. How do your chest, shoulders, rib cage, and belly move as you inhale and exhale? Focus on your body and breath, but don't control the speed or intensity of your breathing.
5. Bring your wandering mind gently back to the body and breath.
6. Do this for two or three minutes at a sitting, and gradually increase the length of time.

DAILY 5×5 PRIORITY: STRETCH AND BALANCE

You'll begin your day with Stretch and close it with Balance. Lions will feel a sense of accomplishment if they bookend their day with positive behaviors. (See Chapter 4 for the exercises or go to MyEnergyQuiz.com for videos.)

* **Stretch.** After waking up at 6:00 a.m. and a quick trip to the bathroom, get on the floor next to your bed and do Child's Pose for one minute, Cat-Cows for one minute, Sphinx for one minute, Dragonfly for one minute, and Cannonball for one minute. By the end of the week, you'll be able to do this progression on autopilot.

* **Exercise.** If you do intense workouts, limit them to three times a week, with a rest period every other day. We recommend Monday, Wednesday, and Saturday at 6:15 a.m.

* **Balance.** As part of your pre-bed winding-down ritual, at 9:30 p.m. do Tree Pose. Stand on one foot for thirty seconds, then switch sides. Keep alternating feet until five minutes have passed. If you lose your balance, just try again. And remember to breathe!

FASTING ENERGY PRIORITY: ESTABLISH AN INTERMITTENT FASTING RHYTHM

We recommend an eight-hour eating window from 10:00 a.m. to 6:00 p.m. All Slow chronotypes have only eight hours to eat, no more. In the first four hours of your sixteen hours of fasting, your body will use up all the carbs you ate that day (but if you move more, you'll burn them faster); then, for twelve hours, you'll be in fat-burning mode. Other benefits: By not eating within three hours of bedtime, your blood sugar rhythm will line up with your melatonin rhythm, and you'll fall asleep easily. Lock into this timing for circadian consistency!

SLEEP HACK: GUAVA LEAF TEA

Along with being antioxidant-rich, guava leaf tea has been found to boost immunity, lower cholesterol, prevent and treat diabetes[6]—and improve sleep quality. One reason Slow Lions might wake up during the night is low blood sugar. That sets off a hormonal cascade that winds up turning on the cortisol tap. Cortisol is one of the circadian hormones that signal our master clock to start a new day, so it's not ideal for it to flow at 3:00 a.m. Guava leaf tea keeps blood sugar stable, so that hormonal cascade never starts, and you stay asleep or calm down enough to fall back to sleep. We recommend the brand GuavaDNA, available online.

JASON'S WEEK ONE ENERGY SCORE: 5

"I did Stretch and Balance as prescribed, and I was ashamed and surprised by how stiff my back is and how hard it was to stay balanced. If you can't stand on one foot for more than five seconds without falling, something inside your body is obviously out of whack. Clearly, I need to improve with nightly practice, which I'm committed to. For the most part, it was all about eating this week, to become more aware of what and how fast I'm chewing. I told my assistant not to bring any food into my office unless I ask for it. Then I turn off my phone and close my laptop, and just eat. No food comes in before ten or after six o'clock. After one week, I swear my pants feel looser. This is fucking awesome."

Week Two

SLEEP ENERGY PRIORITY: POWER-DOWN HOUR

At 9:00 p.m., shut down all screens. Unplug your mind from stressful thoughts or anything that might cause a surge of anxiety. Read a book, listen to music or a podcast, do some easy chores or tasks around the house, talk to a friend, take a bath, have sex! Turn down the temperature of your mind to prepare your body for sleep. Balance poses, accompanied by mindful breathing exercises, will help you become calm and quiet.

DAILY 5×5 PRIORITY: BOUNCE AND BUILD

Along with morning Stretch and pre-bed Balance, this week, add afternoon Build to take advantage of your still-high energy, and evening Bounce to force your body into a second wind when your energy starts to drag.

- **Stretch.** At 6:00 a.m., run through your five-minute progression: Child's Pose for one minute, Cat-Cows for one minute, Sphinx for one minute, Dragonfly for one minute, and finish up with one minute of Cannonball.
- **Build.** At 2:00 p.m., you are already feeling pumped, so channel that power into muscle development. Do one minute of Squats, then a few seconds of rest. Then one minute of Crunches, with some rest. Finish with one minute of Dips.
- **Bounce.** Get a second wind at 7:00 p.m. with one minute of Jumping Jacks followed by a few seconds of rest. Then do one minute of Jumps for Joy, and a short rest. Then two minutes of Skipping (in place or around the house).

- **Balance.** At 9:30 p.m., during your Power-Down Hour, root into your bedroom floor by doing Tree Pose, alternating feet for three minutes. Then try Tippy Toes for two minutes.

FASTING ENERGY PRIORITY: TWO MEALS, ONE SNACK

Any kind of schedule reinforces other rhythms. So having an eating schedule, not just an eating window, will help Slow Lions sleep better, digest with more regularity, and lose weight if that is what they need. Jason used to graze mindlessly all day long. By organizing his eight hours of eating into a concrete schedule of Meal 1, Meal 2, and one or two snacks in between, he winds up eating fewer calories overall and putting less strain and drain on his digestive system.

- **Meal 1:** Eat your first meal of the day at 10:00 a.m., when your eating window opens.
- **Snack 1:** Eat your first snack when the Bears around you have lunch, around 12:00 p.m. to 1:00 p.m.
- **Snack 2:** Have a snack after your 2:00 p.m. Build (this is optional).
- **Meal 2:** Begin eating your second meal no later than 5:30 p.m., and take your last bite at 6:00 p.m.

JASON'S WEEK TWO ENERGY SCORE: 6 (+1 FROM THE PREVIOUS WEEK)

"Big energy bounce this week. I credit the naps, which, once I gave myself permission to take, have been incredible. I lock the door, lie down on my office couch, and I'm out. It's incredible. I haven't napped since I was a little kid, and back then, I fought my mother and teachers every time. Now, these naps are like little escapes, like a condensed trip to Bora Bora. I do Dr. Breus's Nap A Latte trick of drinking coffee before the nap, and waking up when the caffeine kicks in. [See Nap A Latte instructions on page 69.] I've also lost weight by scheduling meals and sticking with intermittent fasting. It's like freaking magic! The weight is just falling off, and I'm not hungry. Very pumped this week!"

Week Three

DAILY 5×5 PRIORITY: SHAKE

The final piece—Shake—is added this week. At 9:00 a.m., after you've been hard at work for a couple of hours, stand up from your desk, Shake yourself up, and boost circulation of blood and oxygen to all the joints that have been locked into a seated posture for too long.

- **Stretch.** At 6:00 a.m., do your spine-limbering flow: Child's Pose for one minute, Cat-Cows for one minute, Sphinx for one minute, Dragonfly for one minute, and finish up with one minute of Cannonball.
- **Exercise.** Stick with a 6:15 a.m. Monday-Wednesday-Saturday schedule.
- **Shake.** At 9:00 a.m., start with one minute of Neck Looseners. Then do thirty seconds of Arm Circles on each side. Move on to thirty seconds of Leg Swings on each side. One minute of Crescent Bends. Finish up with one minute of Trunk Twists.
- **Build.** At 2:00 p.m., do one minute each of Crunches, Dips, Squats, and Wall Sits. If you're feeling your power, add one minute of Kicks.
- **Bounce.** At 7:00 p.m., challenge your coordination and heart by doing two minutes of Burpees, followed by a one-minute rest. Then do two minutes of Ice Skater Jumps.
- **Balance.** At 9:30 p.m., do Tree Pose for five minutes, alternating feet every forty-five seconds.

STACEY SAYS...

It's funny—you would think that because I'm a fitness instructor and personal trainer of thirty years, I would jump out of bed ready to train, run, bike, swim, or do anything at all. Hardly! It's lucky for me that I have a very physical job that keeps me active. When it comes to training my *own* body, though, I struggle with motivation.

There are many, many days when even just the thought of working out makes me tired. I'm fifty-three years old and don't move the way I did when I was in my prime. I'm usually a little fatigued from teaching at SoulCycle, and I personally don't want to spend more time in a gym or a studio. Now, I know I should work out for myself, take stretch, dance, or yoga classes, maybe throw in a few cycling classes, but honestly, I'd rather skip it.

So what do I do to charge myself up physically? Power walking. It's one of the most underrated forms of exercise out there. Walking at a brisk pace, moving my arms (sometimes with three-pound weights depending on my fatigue level), and going hard for about thirty to forty minutes has been one of my favorite motivation hacks ever.

I am known to be seen in Central Park in New York City pounding the pavement on a weekly basis. Not only do I get that time to work my entire body on a muscular-skeletal level; I train my respiratory system, increase circulation, boost my mood, clear my head, and gain an overall sense of balance. If the sky is blue and the sun is out, the vitamin D rush is a bonus.

EMOTIONAL ENERGY PRIORITY: GIVE IT AWAY NOW

Slow Lions, like all slow metabolic types, need to be careful about carrying extra weight, with its exhausting effects and associated diseases like high cholesterol, high blood pressure, diabetes, and certain cancers. Jason is under a lot of stress, is too sedentary, and is a bit too heavy. Along with his Power Protocol energy gains and weight loss, he can use emotional energy to lift himself up, improve his heart health, and improve the lives of others all at the same time.

Researchers from British Columbia studied the effect of giving on cardiovascular health by asking 186 people with high blood pressure how much money they gave to charities.[7] After two years, they checked in with the participants and noted that the most generous among them had lower blood pressure readings than the less generous, regardless of income, education level, and age. Not everyone has money to spare. Giving time and attention also works.

For part two of the same study, the researchers gave 73 hypertensive participants forty dollars three times during a six-week period and asked half to spend the cash on others, and half to spend it on themselves. Sure enough, the give-it-away group had lower blood pressure than their mine-all-mine counterparts. The cardiovascular benefits were on par with adopting a healthier diet and exercise regimen. Per the study, the participants who saw the most improvement gave the money to people they

> Energize Hack: Send a small gift to a friend or family member who's down. Or pick up the check. Or slip a ten-dollar bill into your kid's pocket. You'll see great gains in your health and energy.

loved and cared about. The study authors concluded that charitable giving lowers stress and increases a sense of human connection, both of which are good for your heart — and, we'll add, good for your emotional energy. Bottom line: Giving money away regularly gives *you* invaluable health and energy.

JASON'S WEEK THREE ENERGY SCORE: 7 (+1 FROM THE PREVIOUS WEEK)

"I had a bad week and lost some ground on my program. A work deadline and a health crisis with my elderly mother knocked me off my schedule in just about every category except eating. I didn't stick with my meal timing, but I did stay in my intermittent fasting window. Although my movement took a serious hit this week, I did appreciate the difference when I went back to sitting most of the time. I also noticed that if I don't get that hour to unplug before sleep, I bring my problems into bed with me. I fell asleep okay, but I woke up during the night several times this week with a racing heart. Things will get back to normal next week, and I'll resume the program. It's comforting to know that I have a plan to help me."

Week Four

SLEEP ENERGY PRIORITY: GO A BIT LONGER

How do you feel at 9:00 p.m.? Are you dragging or able to keep going for another full hour? If your goal was to be functional until 10:00 p.m., look at how far you've come. Did you gain fifteen minutes of alertness in the last few weeks? Thirty minutes?

This week, push yourself to stay awake until 10:00 p.m. using all the techniques we've discussed: napping for twenty minutes at 1:00 p.m.; eating on a fixed 10:00 a.m. to 6:00 p.m. schedule; allowing screen exposure until 9:00 p.m. That should get you to 10:00 p.m. Seven and a half hours of uninterrupted sleep later, you'll wake up fully charged and restored.

DAILY 5×5 ENERGY PRIORITY: GO HARDER

This week, up the intensity.

- **Experiment.** Check the movement list in Chapter 4. Are there any you have been afraid of trying, like Eagle, Dancer, or Gorilla Burpees? This is the week to face your fears and experiment with the movements that you've avoided.
- **Intensity.** Level up. If you have been coasting with Easy level, go Medium. If you've been doing Medium, do Hard. If you've been doing Hard level, up your amplitude. Jump higher, Skip faster, Squat deeper.
- **Duration.** Extend your Daily 5×5 to a Daily 6×5, or a Daily 5×7. Add a session or minutes if you are inspired.
- **Frequency.** If you've only been doing three sessions a day—some days are just like that—really push this week to hit the magic number of five.

FASTING ENERGY PRIORITY: SYSTEMS CHECK

After three weeks of intermittent fasting, you have already adapted to the change in your eating rhythm and have probably noticed that your clothes are a bit looser. This week push your window by fifteen minutes on either end to test your endurance. If you just *can't*, okay. But you won't know unless you try.

- **Open** your eating window at 10:15 a.m.
- **Close** it at 5:45 p.m.

JASON'S WEEK FOUR ENERGY SCORE: 8 (+3 SINCE WEEK ONE)

When he started his Power Protocol, Jason had several goals:

- **Level up his baseline energy.** "I've done that, pretty easily. From where I started to where I am now, I feel like a new man. I think I overestimated my physical energy when I started this. Looking back, there was no reason I should have felt that drained when I fell into bed. I exhausted my body. Now, I energize."

- **Change his movement and eating to reduce body aches and his belt size.** "Well, I've lost 7 pounds and bought new pants," he said. "The most improved movement category is Balance. When I started, I would fall out of Tree almost immediately. Now I can hold it on one foot for three minutes straight. That one move taught me the importance of balance in all my ways. Balance eating with fasting, activity with rest, excitement with quiet time."

- **Lower his blood pressure, blood sugar, and cholesterol.** "The doctor's visit awaits to check my cholesterol, but I have an at-home blood pressure monitor and blood glucose test kit, and my numbers are definitely getting better. The longer I do this, the better they'll be. Four weeks got me started, but I'm going to keep going!"

Another life energized. Our work here is done.

TROUBLESHOOTING

Slow Lions, a few things to watch out for:

1. **Impatience.** Slow types have the most to gain from this program, but it will take commitment. Your metabolism is like a cruise ship; you can spin the wheel, but it takes some time to change direction. We guarantee that you will lose weight and feel supercharged. Give it two weeks.

2. **A bumpy adjustment.** Lions are at the mercy of their efficiency. It's not easy for them to change their routines or try new things. If you expect to feel out of sorts as you adjust, you'll make a smoother transition, and the extra energy will be worth it.

3. **Overlooking energy sources.** You might get so caught up in the physical energy gains, you forget about the emotional ones. We're helping you find more hours in the day so you can spend them with friends and family to make powerful (and empowering) human connections. Take advantage of emotional energy opportunities.

Daily Schedule

The end of Week Four is the beginning of the rest of your high-energy life. Stick with it, and you'll always feel the power.

6:00 a.m. Wake up. Next-level movements — double time optional: 1 minute Child's Pose; 1 minute Cat-Cows; 1 minute Sphinx; 1 minute Dragonfly; 1 minute Cannonball; drink 500 milliliters of water

6:15 a.m. Exercise (optional) 45 minutes to 1 hour, Monday, Wednesday, Saturday

9:00 a.m. Next-level movements — double time optional: 1 minute Neck Looseners; a few seconds rest; 1 minute Arm Circles; a few seconds rest; 1 minute Leg Swings; a few seconds rest; 1 minute Crescent Bends; a few seconds rest; 1 minute Trunk Twists; drink 250 milliliters of water

10:15 a.m. Eating window opens

1:00 p.m. Power nap (optional)

2:00 p.m. Next-level movements — double time optional: 1 minute Squats; a few seconds rest; 1 minute Crunches; a few seconds rest; 1 minute Dips; a few seconds rest; 1 minute Wall Sits; a few seconds rest; 1 minute Kicks; drink 250 milliliters of water

5:45 p.m. Eating window closes

7:00 p.m. Next-level movements — double time optional: 2 minutes Burpees; 1 minute rest; 2 minutes Ice Skater Jumps; drink 250 milliliters of water

9:00 p.m. Power-Down Hour; turn off all screens

9:30 p.m. 2 minutes Dancer; 2 minutes Eagle; 1 minute Figure Four

10:00 p.m. Bedtime

CHAPTER 13

The Fast and Medium Dolphin Power Protocol

Leigh is a forty-five-year-old mom in the suburbs of Chicago who works at home full-time. She's a step ahead on the Energize protocol because she's been intermittent fasting for a while already. "My eating window is eight hours, starting at 10:00 and ending at 6:00. It's not great for social-izing, but it works for me. My husband is a Bear, and he loves nothing more than meeting up with people and drinking all evening. Our friends have gotten used to me sitting at the table and sipping tea while they all eat a full meal and have cocktails. My main complaint, which led me to read the Sleep Doctor's blog and get in touch with him, is insomnia. I've learned that when I do eat dinner late, I can't sleep—even with a pill. I feel overly full and my stomach gets upset. Recently, I was pressured to have an appetizer and a glass of wine around 9:00 p.m., and, as I pre-dicted, I was sick to my stomach and couldn't sleep. It's a hard rule for me: If I eat after 6:00 p.m., I know I'm going to have a bad night, which spirals into a bad day, or a bad few days."

As for movement, Leigh is parked at her desk for most of the day. "A typical day starts at 6:30, when I wake up. I sit at my desk for a couple hours. Then I do some morning chores and eventually have tea and breakfast at 10:00. I'm back at my desk for five hours straight to meet my first daily deadline. When I hit 'send' on that, I get up, pet the cats, and make some lunch. I take it back upstairs to eat at my desk. I don't get back up until dinnertime." That adds up to eleven hours of chair time, with two short breaks for food. "Well, it's not that I'm a stone statue at my desk. I'm a fidgety person. I have an uncomfortable desk chair and I'm constantly changing position and trying to get comfortable. I burn a lot of calories that way."

She exercises with a trainer twice a week "without fail," she said. "My slot is 7:00 p.m. I'm much more coordinated in the evening than in the morning. When I've tried to do morning workouts, it's ridiculous. Even my trainer has noticed."

Leigh (wrongly) considers it a personal failing that she has insomnia. "I've never been able to just fall asleep, and I always blamed myself," she said. "I remember reading the book *Go, Dog. Go!* as a kid. One picture is a group of dogs, all of them asleep, and one dog's eyes are wide open with exhausted circles drawn around them. I thought, 'I am that dog.' Why can't I be like the other dogs? I use sleep medication more than I should. I know that if I slept more than five or six hours a night, I'd have more energy. But I can't do it."

Leigh's energy goals:

- To sleep more than six hours per night.
- To try to taper off medication.
- To move more throughout the day.

WATCH OUT FOR THE BIGGEST ENERGY DRAINS FOR FAST AND MEDIUM DOLPHINS

Many of Leigh's behaviors are sucking her energy dry.

⬇ Energy Drain: To Pill or Not to Pill, That Is the Question

When you are staring at the ceiling for hours at a time, the promised relief of just taking a pill to fall asleep is so very tempting. But the short-term reward of hypnotic drugs doesn't come close to outweighing the long-term risks. To be clear, we feel strongly that this is a discussion you should have with your prescribing health care provider. There are many people who legitimately need a pill to sleep, and that is "healthy" in many situations (severe mental health disorders, anxiety disorders, depression, pain, etc.).

For everyone else, unfortunately, the data about using sleep aids is not that great:

- A large-scale analysis found that users only get an additional eleven minutes of sleep per night.[1]
- The all-natural techniques to get to sleep—like progressive muscle relaxation and cognitive behavioral therapy—are actually more effective than meds[2] and have none of the risks, like waking up with brain fog.

↑ Energy Gain: Not to Pill, That Is the Answer

Prescription and over-the-counter sleep drugs can help people break the cycle of sleeplessness. If it's a question of no sleep for days on end or taking a pill, take the pill. For long-term solutions to chronic insomnia, the American Academy of Sleep Medicine, the World Health Organization, and most physicians agree: Drugs are not the best medicine.

There is research that occasional use of valerian root, melatonin, and CBD supplements can be effective in breaking the insomnia cycle. (More details in Chapter 3.) However, if you require some kind of pill every single night, see a professional to explore whether a mental or psychological issue, high stress, or chronic pain is interfering with your sleep. Popping a pill is an easy way to address the issue, but then you can become dependent. A better solution might take some effort, but real answers usually do, and they will be more energizing in the end.

↓ Energy Drain: Feeling Guilty (Your Mom Inflicts Enough of That Already)

Michael has observed that, for his patients who find themselves lying awake and ruminating, their negative thoughts are often self-directed. The first moment they have for any kind of introspection all day might be when they get in bed at night. The light goes off, and the dark and dark thoughts come flooding in. For Fast and Medium Dolphins, who are full of nervous energy when the sun goes down, self-blame and shame about everything, not just their nightly struggle to sleep, might keep them up for hours.

Leigh said she feels like a failure for not being able to do the one thing that seems to come so easily and naturally for "everyone" else. But insomniac Dolphins' chronorhythm is fundamentally different from the other three chronorhythms. The body temperature and blood pressure of Bears, Wolves, and Lions drop before bed; Dolphins' core temperature and blood pressure go *up*. Their energetic arousal rhythm — feeling awake and alert — is on the upswing when the rest of the world is winding down.

While others have a gently curving cortisol rhythm, Dolphins' rhythm is a steady *incline*, making them increasingly fidgety and anxious as the day wears on. Combine type A Dolphin traits with a speedy Fast or Medium metabolism, and you have all-day edginess plus chronic exhaustion.

The phrase "the weight of a guilty conscience" comes to mind. In fact, researchers from Princeton University have studied whether that particular emotion really does weigh us down and make going through everyday motions an exhausting slog.[3] Can abstract metaphors translate into bodily experience? The scientists reviewed four studies to see if "the embodiment of guilt" is real. In one study, a group of participants were asked to remember and describe in detail a time they committed an unethical act. Compared to the control group, the first group subjectively estimated their body weight to be heavier than it was before they had the guilt-inducing memory. In another study, participants were asked to recall an unethical experience in their past, and to report on the perceived effort to do a minor task, like carry groceries. The participants who were weighed down by guilt-inducing thoughts perceived it would take more effort to do these minimal physical tasks than the control group did.

Guilt is literally an exhausting drag on our bodies.

↑ Energy Gain: It's Safe to Go Cold Turkey on Guilt

The cure for self-blame is self-acceptance. Fast or Medium Dolphins have done nothing to cause their insomnia. You didn't ask to be an insomniac. Your genes gave you a low sleep drive. You were born this way, and you can't change who you are. However, *we have your back.* When you can accept your chronotype, you can begin to work with what you've got by following your Power Protocol and practicing the techniques in this chapter, and you will then have better sleep and more energy.

↓ Energy Drain: Working Out Occasionally

Leigh described her twice-weekly trainer sessions as an excellent release value for her constant stress and pent-up nervous energy. Fitness is fantastic, for one and all, and we're impressed whenever we hear about a Dolphin being a regular exerciser. For the most part, insomniacs tend not to be obsessive about fitness or healthy eating. If fitness "fits" into your schedule, great. But it can also become an exhausting, anxiety-and-guilt-producing cross to bear when you don't have the time or will to do it.

> 💥 Dolphin Hack: The five daily movements are enough to push the emergency release button on the pressure cooker that is your body.

↑ Energy Gain: Surgical Detachment from the Chair

Consistent daily movement vs. twice-weekly workouts: The former is like riding a long, powerful wave with control and joy all week long. The latter is like riding two monster waves and otherwise falling asleep on your surfboard.

The award for "most fidgety" goes to Fast and Medium Dolphins. Congratulations! This is one of the reasons you maintain an average to low BMI.[4] But weight maintenance is not the same as feeling energized. Even regular-exerciser Dolphins like Leigh report feeling tired most of the time. The objective of our program is to give you abundant power to *go* and *do,* and that takes a bit more conscious effort than mindless fidgeting. So, keep squirming in your chairs, but also set your alarms and make a habit of purposefully increasing your muscle mass, improving circulation, and making the heart beat a bit faster several times a day.

SLEEP HACK: GUAVA LEAF TEA

Along with being antioxidant-rich, guava leaf tea has been found to boost immunity, lower cholesterol, prevent and treat diabetes[5]—and improve sleep quality. One reason Dolphins might wake up during the night is low blood sugar. That sets off a hormonal cascade that winds up turning on the cortisol tap. Cortisol is one of the circadian hormones that signal our master clock to start a new day, so it's not ideal for it to flow at 3:00 a.m. Guava leaf tea keeps blood sugar stable, so that hormonal cascade never starts, and you stay asleep or calm down enough to fall back to sleep. We recommend the brand GuavaDNA, available online.

YOUR NEW KICK ASS ALL DAY, EVERY DAY PLANNER

Program these alarms on your phone for these key times:

7:00 a.m. Wake up; Stretch, plus water

9:00 a.m. Eating window opens

10:00 a.m. Bounce, plus water

2:00 p.m. Build, plus water

4:00 p.m. Bounce, round two, plus water

6:00 p.m. Exercise (optional)

7:00 p.m. Eating window closes

9:00 p.m. Shake, plus water

11:00 p.m. Power-Down Hour; turn off all screens

11:30 p.m. Balance (no water)

12:00 a.m. Bedtime

Week One

Begin on Monday. Any Monday. THIS Monday.

SLEEP ENERGY PRIORITY: CONSISTENT WAKE TIME AND BEDTIME

You have only two sleep energy tasks this week. *Make sure you get out of bed at 7:00 a.m. and get into it at midnight.* You will try to sleep only during those seven hours per day. If you stick with this plan for a week, you should see improvement in how long it takes you to fall asleep and how long you stay down.[6]

> ✴ Energize Hack: By strictly limiting in-bed hours, you train your body to seize the opportunity for sleep when it's presented to you. We call this *brain training*!

DAILY 5×5 PRIORITY: DOUBLE BOUNCE

To counteract their fatigue, Fast and Medium Dolphins need to jump up and down, get their hearts pumping, and increase oxygen circulation to every cell in their body. We recommend double the Bouncing fun for your type. The morning Bounce helps break through brain fog, and the afternoon Bounce expends some of your nervous energy through movement so you can concentrate on the task at hand. (See Chapter 4 for the exercises or go to MyEnergyQuiz.com for videos.)

- **Morning Bounce.** At 10:00 a.m., do two minutes of Jumping Jacks, followed by a one-minute rest. Then do another two minutes of Jumping Jacks.
- **Afternoon Bounce.** At 4:00 p.m., do two minutes of Jumps for Joy, followed by one minute of rest. Then do another two minutes of Jumps for Joy.

FASTING ENERGY PRIORITY: ESTABLISH AN INTERMITTENT FASTING RHYTHM

We recommend the ten-hour eating window from 9:00 a.m. to 7:00 p.m. You'll have only ten hours to eat, no more, and you'll stop eating five hours before the get-in-bed time of midnight, preventing any gastrointestinal issues from distracting you as you're trying to fall asleep.

LEIGH'S WEEK ONE ENERGY SCORE: 3.3

"I spoke to my doctor, who told me it might take three to six weeks to taper off sleeping pills. I had no idea that it'd take so long. We mapped out a plan. Starting next week, I will cut my dosage by 25 percent per night per my doctor's orders. I like the security of still taking a pill, while also knowing I'm doing something to get myself off them eventually. I've been sticking with the seven hours in-bed timing, but I'm probably sleeping only five hours a night. I did Bounce about half the time and liked it when I did. I'm going to make an effort to do more next week."

Week Two

SLEEP ENERGY PRIORITY: THE BED IS NOT A BATTLEFIELD

"Stimulus-control therapy" is a technique that gets you to stop thinking about your bed as if it were a torture device and to learn to associate it with sleep. Decades of scientific research have found the technique to be effective and reliable for the treatment of insomnia.[7] No more reading, watching TV, scrolling on FB, eating, talking, or working on that particular piece of furniture. Associating your bed with other activities perpetuates insomnia. Insomniacs tend to see it as a battlefield, where they go to suffer and struggle. To make peace with your bed, only use it for sleep and sex; cuddling is good, too.

DAILY 5×5 PRIORITY: STRETCH AND BALANCE

This week, add the morning and nighttime sessions to your day. Within a few minutes of rising at 7:00 a.m., take a quick trip to the bathroom to pee, and then get on the floor of your bedroom to Stretch for five minutes and ignite the pilot light of your day. At 11:30 p.m., Balance to focus the mind on anything but the threat of insomnia.

- **Stretch.** At 7:00 a.m., start with one minute of Child's Pose, then move into a minute of Cat-Cows. Do a minute of Sphinx, a minute of Dragonfly, and finish up with a minute of Cannonball.

- **Morning Bounce.** At 10:00 a.m., spend five minutes learning and mastering Burpees.
- **Afternoon Bounce.** Spend five minutes at 4:00 p.m. doing Ice Skater Jumps.
- **Balance.** At 11:30 p.m., do five minutes of Tree Pose. First stand on one foot for thirty seconds, then switch sides. Alternate feet until the time has passed.

FASTING ENERGY PRIORITY: SCHEDULING CAFFEINE

Drinking coffee and caffeinated tea, soda, and energy drinks is like walking a tightrope for Fast and Medium Dolphins. You might reach for those beverages in the morning to clear brain fog—which doesn't actually work; try fifteen minutes of sunlight instead. That won't affect your sleep later on. But if you have any caffeine after 3:00 p.m., it'll sabotage all your efforts to teach your body to go down during those seven in-bed hours. We recommend that Dolphins have coffee or tea—just one 250 milliliters serving—with the midday meal. Skip it in the morning so you don't get jittery (as if you need more of that). Use an all-natural method to calm down, like deep breathing (see instructions on page 253) or sex (no instructions needed). And don't have any caffeine later in the afternoon or after dinner, ever. *Ever.*

> Energize Hack: Only have one cup of caffeinated beverage per day, with a pre–3:00 p.m. meal.

LEIGH'S WEEK TWO ENERGY SCORE: 5 (+1.7 FROM THE PREVIOUS WEEK)

"My sleep-med dosage dropped again, but no withdrawal symptoms, thank God. I think the new movement schedule is helping. I'm getting about six hours a night on average. On Wednesday night, I got in bed at midnight, passed out within half an hour, and didn't wake until my usual time of 6:30 a.m. Rumination is way down, simply because I spend so much less time in bed. My body is learning. I miss my all-day tea drinking, though. I've switched to decaffeinated tea, which is like drinking hot potpourri. But I'll get used to it. I like Balance at night, but I'm not crazy about Stretching. This is all so new to me; I just have to keep at it to see the benefits. Once things really start to 'work,' I'm sure I'll feel more motivated. I hate Burpees, by the way. My trainer has made me do them for years and I always complain. Now I'm making myself do them! If she only knew ... But to be clear, I still hate them."

Week Three

SLEEP ENERGY PRIORITY: POWER-DOWN HOUR

If your brain is running at a certain pace all day long, it doesn't just stop the second you want to fall asleep, especially for Dolphins. Your upside-down biology makes you feel most awake when it's time for sleep. It's not your fault, but you can do something about it.

The winding-down objective is to lower your blood pressure and body temperature to make a smoother transition from your energetic peak in the late evening into rest. For the other chronotypes, the process of vasodilation—blood leaving the body core and heading to the extremities—warms the feet and hands but lowers the body temperature, which coincides with the release of melatonin. If your feet are always cold in bed, it might mean your vasodilation circadian rhythm is screwed up, a common problem for insomniacs.[8]

It might seem counterintuitive to take a hot bath or shower before bed when you need to *lower* your core temperature. But when you get out of the tub, your body temperature drops. Forcing that drop is a biohack that flips on the sleep switch.[9] Another trick is to lower the room temperature to the National Sleep Foundation's recommendation of 60 to 67 degrees Fahrenheit, so crank the AC or open a window an hour before bedtime to chill the space.

Lower the temperature of your mind, too, by getting off all devices. Your fast, agile brain needs to disengage well before bedtime, or your thoughts will keep racing. Stop reading stimulating emails, status updates, tweets, articles, and blogs. The melatonin-suppressing blue light from an iPhone two inches from your face doesn't help either. Turn it all off. Don't worry. Any news that breaks before morning will still be there when you wake up.

Other recommendations for winding down: organize your desk, pack lunches for school, get into your pj's, pray, or try progressive muscle relaxation or meditation. Just unplug.

DAILY 5×5 PRIORITY: BUILD AND SHAKE

The last two sets to incorporate into your Daily Schedule this week are Build and Shake. You'll Build muscle mass at 2:00 p.m. to increase your power as you enter the late afternoon. At 9:00 p.m., it's all about Shaking off stress and stiffness so you can relax.

- **Stretch.** At 7:00 a.m., run through your circuit of all five spine stretches: one minute of Child's Pose, one minute of Cat-Cows, one minute of Sphinx, one minute of Dragonfly, and one minute of Cannonball.
- **Morning Bounce.** At 10:00 a.m., Skip around the block for five minutes. If you can't go out, Skip in place indoors.
- **Build.** At 2:00 p.m., stop what you're doing and do Squats for one minute. Then do a minute of Crunches. Rest for a minute, and repeat.
- **Afternoon Bounce.** At 4:00 p.m., do five minutes of your jump of choice.
- **Shake.** When your energy goes up at 9:00 p.m., do one minute each of Neck Looseners, Arm Circles, Leg Swings, Crescent Bends, and Trunk Twists, with short rests in between.
- **Balance.** At 11:30 p.m., keep it grounded with three minutes of Tree Pose and two minutes of Tippy Toes.

EMOTIONAL ENERGY PRIORITY: MEDITATION

It's not your fault you have insomnia. Your brain kicks into high gear when everyone else is shifting into neutral. Leigh said, "Sometimes, I glance in the mirror and think, 'My god, I look exhausted. I would look ten years younger if I could just go to sleep like a normal person. What's wrong with me?'" Nothing is wrong with you, Dolphins. You are different, that's all. And being different is *good*. You dance to your own rhythm, and your mind is always busy coming up with new ideas and projects.

We recommend a two- or three-minute session of diaphragmatic breathing—deep belly breaths—during your Power-Down Hour to help prepare your body for sleep and give you a positive emotional boost. According to a wealth of research, this kind of meditation improves focus, raises positive affect, and lowers cortisol levels.[10] It also stimulates the vagus nerve, which runs from your neck down to your guts. This nerve turns on the switch for the rest-and-digest parasympathetic nervous system and turns off the fight-or-flight sympathetic nervous system. Since your fight-or-flight hormones go up at night, this is the perfect emotionally energizing exercise for you.

↑ Calming Diaphragmatic Breathing for Beginners

1. Sit in a comfortable chair or lie flat on the floor on a mat.
2. Place one hand on your belly just beneath your ribs. Place the other hand on your sternum.
3. Inhale deeply through the nose into your belly until your hand rises for a count of four. Your chest should remain still.
4. Hold for four.
5. Exhale through pursed lips and let your belly and hand deflate for four. Keep exhaling until all the air is expelled.
6. Repeat ten times. Keep it slow and steady.

THE ULTIMATE DOLPHIN GADGET

Since you, Dolphins, are detail-oriented and have highly active minds, you might want to make an investment in a guided meditation / sleep tracking headband called Muse (ChooseMuse.com). The gadget is an EEG brain-sensing device that tracks your mental activity. It has sensors that go over your forehead and behind your ears, and it connects to a phone app. When the EEG picks up chaotic brain activity—your default setting, Dolphins—it plays storm sounds. When your brain is calm, it plays peaceful sounds, like gentle rain or wind. This is very useful when meditating. You'll know instantly when your mind starts drifting and you need to return your focus back to your breath. When you hit that calm zone, the sounds will change to reflect that. After a meditation session, you can check the app's graph, which shows your changes in brain activity. With this data, you can willfully, mindfully improve your focus over time.

LEIGH'S WEEK THREE ENERGY SCORE: 6 (+1 FROM THE PREVIOUS WEEK)

"Getting there. I really do feel better this week. I'm doing about half the movement schedule, which is still more than I was doing before. It's just not always possible to stop whatever I'm doing to jump up and down. But when I can, it does feel good and lifts me up. The secret sauce to this whole program is having little things to do every few hours that are for me and my wellness. I've never spent so much time before working toward feeling better, and now that I am, I get what it's all about. It feels kind of good to take care of myself for a change. I think I am starting to get a little of that self-acceptance Stacey and Michael have been talking about. My day doesn't have to be a long slog from responsibility to responsibility. If you do what you know feels good many times a day—like stretching and jumping—you'll have more energy."

Week Four

SLEEP ENERGY PRIORITY: TRY, TRY AGAIN

The thing about insomnia: Even when you follow the rules, bad nights will always happen, usually at the worst possible time. No one has perfect sleep, *nobody,* not even the Sleep Doctor himself. Even if you practice the techniques we've described, you'll occasionally find yourself lying awake, feeling like a failure at this basic human function. It's annoying and exhausting not to fall asleep. But it's not the end of the world.

You are not your insomnia. It's a side effect of being you, but it doesn't define you. Stop self-blame by practicing a self-compassion exercise. These two come from Kristin Neff, PhD, author of *The Mindful Self-Compassion Workbook: A Proven Way to Accept Yourself, Build Inner Strength, and Thrive.*

Exercise 1: How would you treat a friend? If your best friend were in the exact same boat as you and told you that she hated herself because of some body function that she was born with, what would you say to her? Would you tell her to blame herself and feel like a failure? Of course not! You'd encourage her to try some evidence-based strategies and do her best, and not to fault herself for her DNA. Biohack alert: The "friend perspective" forces your brain activity out of the amygdala's emotional center and into the prefrontal cortex, the logical, rational center. When you take the emotion out of the conversation, you can take control of your thoughts.

Exercise 2: Quiet your critical self-talk. According to Dr. Neff, the first step is to clue in to self-bashing when it happens. For you, it's probably while you're lying awake at night, ruminating and beating yourself up for not falling asleep. When you hear yourself doing it, reframe your thoughts toward compassion. For example, the usual self-talk might sound something like this: "What's wrong with you? Why can't you do this one thing that even babies can do effortlessly?" The friendly reframe: "You're an amazing wife/husband/mom/dad/sister/brother/daughter/son. You work hard and care deeply about others. You have so many gifts. Unfortunately, you do struggle to sleep. But that's just a tiny glitch in an otherwise lovable you."

This cognitive shift doesn't happen overnight. You might need to practice a nightly self-compassion exercise for weeks or months to internalize that a bad night isn't a personal failure. You may need therapy. But you will get there. You will sleep eventually. And every hour is an energizing gift. Appreciate the sleep you get. Tomorrow is another night.

DAILY 5×5 ENERGY PRIORITY: GO HARDER

This week, up the intensity.

- **Experiment.** Check the movement list in Chapter 4. Are there any you have been afraid of trying, like Dancer Pose or Gorilla Burpees? This is the week to face your fears and experiment with the movements that you've avoided.

- **Intensity.** Level up. If you have been coasting with Easy level, go Medium. If you've been doing Medium, do Hard. If you've been doing Hard level, up your amplitude. Jump higher, Skip faster, Squat deeper.

- **Duration.** Extend your Daily 5×5 to a Daily 6×5 or a Daily 5×7. Add a session or minutes if you are inspired.

- **Frequency.** If you've only been doing three sessions a day—some days are just like that—really push this week to hit the magic number of five.

LEIGH'S WEEK FOUR ENERGY SCORE: 7 (+3.7 SINCE WEEK ONE)

When Leigh started the Energize program, she had three goals:

- **To sleep more than six hours per night.** "I've been getting six hours most of the time. Because of that, I've noticed a slow but steady energy increase all month. I still yawn and feel drowsy every day, but much less than before. I'm very excited about continuing with the program to see if my sleep improves and my energy continues to go up, or if I hit a ceiling."

- **To try to taper off medication.** "This has been a success. I like having a systematic approach, and my tapering plan is very detailed, just like the Energize program. I do best when I know exactly what to do. I've reduced my medication by 75 percent. I'm on such a low dose now that I doubt it's doing anything, but it might be a while before I can completely let go of the emotional comfort of taking a pill, even if I've cut it to a speck."

- **To move more throughout the day.** "Definitely moving more! And it's been very positive. My trainer has said that I'm more fit than ever, thanks to my additional Burpees. My weight is at a ten-year low, which makes me feel great about getting dressed each day. Overall, my experience with the Power Protocol has been excellent. Now if I can only get my husband to try Slow Bear, we'll be a fully energized couple."

Another life energized. Our work here is done.

TROUBLESHOOTING

Fast and Medium Dolphins, a few things to watch out for:

1. **Unrealistic expectations.** Insomniacs need to lose the illusion that, on any protocol, their sleep problems will disappear. It is not in a Dolphin's biology to get eight consecutive hours of deep sleep night after night. We can get you to six regularly; that's a win.
2. **Inconsistency.** Energizing comes into full force with *consistency*. Use your superpower of attention to detail by following the protocol as closely as possible, and you'll reap huge benefits very quickly.
3. **Frustration.** Some of the movements will feel awkward until you practice them enough. Ice Skater Jumps are hard to master. But who cares if you feel like a clod? Perfection is not the point. Silence the Olympic judges in your mind and give yourself top marks for effort.

Daily Schedule

The end of Week Four is the beginning of the rest of your high-energy life. Stick with it, and you'll always feel the power.

7:00 a.m. Wake up. 1 minute Child's Pose; 1 minute Cat-Cows; 1 minute Sphinx; 1 minute Dragonfly; 1 minute Cannonball; drink 500 milliliters of water

9:00 a.m. Eating window opens

10:00 a.m. 5 minutes Medium or Hard level jumps of your choice; drink 250 milliliters of water

2:00 p.m. Level up movements—double time optional: 1 minute Squats; a few seconds rest; 1 minute Crunches; a few seconds rest; 1 minute Wall Sits; a few seconds rest; 1 minute Dips; a few seconds rest; 1 minute Kicks; drink 250 milliliters of water

4:00 p.m. 5 minutes Medium or Hard level jumps of your choice; drink 250 milliliters of water

6:00 p.m. Exercise (optional)

7:00 p.m. Eating window closes

9:00 p.m. Level up movements—double time optional: 1 minute Neck Looseners; a few seconds rest; 1 minute Arm Circles; a few seconds rest; 1 minute Leg Swings; a few seconds rest; 1 minute Crescent Bends; a few seconds rest; 1 minute Trunk Twists; drink 250 milliliters of water

11:00 p.m. Power-Down Hour; turn off all screens

11:30 p.m. 2 minutes Dancer; 2 minutes Eagle; 1 minute Figure Four

12:00 a.m. Bedtime

CHAPTER 14

The Slow Dolphin Power Protocol

Marc, a sixty-one-year-old security guard in Los Angeles, has never had any kind of regular fitness or movement routine. "That's not true. In the eighth grade, I had gym class three times a week," he said. "And I hated it." Marc's wife is a yoga person who has tried for years to get him to do just a few poses. But he felt awkward doing Downward Dog and didn't see the point. "A friend told me that exercise only lengthens your life by the amount of time you spend exercising," he said. "It was a joke, but it made sense to me. I don't look good in Lycra, either."

A lifelong insomniac, Marc can't remember ever having an "easy" night's rest. "I've always spent at least two hours wide-awake in bed before I fall into a light sleep. And I wake up earlier than anyone I've ever lived with. I get around five or six hours a night. Since I've gotten older, those sleep hours are interrupted by trips to the bathroom to pee. I've just accepted that it's my lot to always feel tired." When Marc sits down on a chair or the couch to read, he finds himself nodding off. "I take little cat-naps a lot," he said. "I can't help it. As soon as I'm still, my eyelids get heavy."

To combat his constant fatigue, Marc relies heavily on caffeine to perk up his brain and body. He drinks a couple of cups of coffee before his noon to 8:00 p.m. shift and has another during his dinner break. "I eat when I feel hungry, when others are eating, or if I get a craving," he said. "My weight has always been fairly stable, between 10 and 20 pounds over-weight. I've never been thin. I've never been fat-fat. And I've never really cared that much about having a great body. That whole business just seems like a waste of energy. My energy is low enough already."

Marc's energy goals:

- To stop feeling so tired during the day.
- Not to have to rely so heavily on coffee.
- To sleep better at night.
- "My wife wants me to be more active, too, but that's her priority, not mine!"

WATCH OUT FOR THE BIGGEST ENERGY DRAINS FOR SLOW DOLPHINS

Many of Marc's behaviors are sucking his energy dry.

⬇ Energy Drain: Catnaps Are Great for CATS, Not Dolphins!

All Dolphins have erratic sleep schedules that cause sleep deprivation. Slow Dolphins—unlike their type A fidgety Fast and Medium counterparts—are capable of napping during the day. Marc found himself dozing whether he intended to or not. He's like Leonardo da Vinci, the Italian Renaissance artist, sculptor, and inventor, who famously rejected the concept of sleeping at night and, instead, took twenty-minute naps every four hours throughout the day. His grand total of daily sleep only amounted to two hours because of this unconventional schedule. Da Vinci once said, "Sleep resembles death." That opinion, as well as his nap-py lifestyle, makes us suspect that Da Vinci was a severe insomniac who, like Marc, struggled to fall asleep at night and was incapable of staying down for a "normal" seven to eight hours straight. If Da Vinci and any doze-dependent Slow Dolphins stopped taking those catnaps, they might be able to get closer to six hours of continuous sleep overnight.

Given Dolphins' low sleep need (compared to Bears and Lions, who need their eight hours), six continuous hours is enough rest and restoration for them to *go* and *do* the next day. But to get that unconscious time, they have to build up sleep pressure so that, by midnight, they feel tired enough to go down without too much anxiety. The longer the nap, the worse off they'll be come bedtime.

↑ Energy Gain: If You're Moving, You're Not Dozing

The fix for unintentional or intentional daytime dozing is to get up and move every few hours. The Daily 5×5 plans bursts of increasing your heart rate and oxygen intake throughout the day, boosting energy and chasing away fatigue. If you're jumping and stretching, you won't be on the couch, napping.

↓ Energy Drain: The Exhaustion Cycle: Inactivity, Insomnia, Repeat

Sleep in its most basic sense is *recovery,* but you must have something to recover from! Being sedentary brings on poor sleep, which leads to feeling too exhausted to move the next day, and so it goes:

Insomnia → exhaustion → not moving → insomnia →
exhaustion → not moving

It's a bad pattern that is all too common for Slow Dolphins. Once you get locked into it, it might feel nearly impossible to escape. It's like a really ugly pattern of wallpaper that repeats forever...until you tear it down and replace it with something much prettier.

↑ Energy Gain: Fake It Till You Make It

You hate exercise, we get it. But what we recommend isn't exercise; it's movement. You are basically doing it anyway. Just do it at the right times and—guess what—you will sleep and feel better. Don't just believe us. According to a recent study by researchers at the Yale School of Nursing, there is a causal relationship between daytime physical activity and longer-duration sleep.[1] The insomniac study participants wore monitors to track their daily movement and sleep over a fifteen-day period. It was found that when they moved more during the day, they slept longer at night; if they slept longer at night, they moved more the *next* day.

Slow Dolphins face a very clear choice between their bad pattern (exhaustion, inactivity, insomnia) and a bold new one (movement, energy, decent sleep).

⚡Energize Tip: What might hold you back from making positive change is habit. We take comfort in the familiar, even if it's bad for us.

You might resist moving more because your energy is so low to start, and just the idea of five movement sessions is exhausting. Motivate yourself with the mantra "Nothing changes if nothing changes." Or "I can do *anything* for five minutes." The entire first week on this protocol, you only have to do *one* movement session per day. Just one. We're only asking for five minutes per day. You *know* you can do that much.*

↓ Energy Drain: Snacking Is Attacking (Your Sleep)

As we've mentioned, there are dozens of internal body clocks that run on circadian rhythms. You have a gut clock that tells you when to feel hungry and when to stop eating, just as you have a master clock in your brain that tells you when to feel tired and when to feel alert. Granted, for Dolphins, your master clock has an unusual rhythm. But you can boost it by syncing up your metabolic clock, the one that controls the rise and fall of your blood sugar.

As you know, blood sugar goes up and down depending on what you eat and when you exercise. But it also has a rhythm of its own. Blood sugar is an ingredient in the circadian cocktail (along with cortisol and adrenaline) that's released two hours before wake time. A natural decrease in blood sugar at night allows you to slow down and feel tired.

Researchers in the UK set out to test how mealtime impacted our master clock's melatonin and cortisol rhythms by strictly regulating the eating schedules of their healthy male participants over a thirteen-day period. They discovered that late meals delayed blood sugar from dropping by five hours. For patients with circadian-rhythm disorders,[2] like all insomniacs, this delay signals to your body that it's not time for rest yet.

---

* For more on establishing positive routines, check out *Tiny Habits: The Small Changes That Change Everything*, by BJ Fogg.

↑ Energy Gain: No More Midnight Meals

Get your blood sugar rhythm back in sync by not eating for at least six hours before your midnight bedtime. We're recommending an eight-hour eating window of 10:00 a.m. to 6:00 p.m. Eating during daylight hours only is a zeitgeber, a circadian reinforcer. Sixteen hours of fasting between 6:00 p.m. and 10:00 a.m. will give you approximately twelve hours in the fat-burning zone, which can help Slow Dolphins drop excess pounds.

What's more, if you eat on this schedule, you'll be less sleep-deprived and will make healthier food choices. A recent study found that insomniacs who are exhausted give themselves high-sugar, high-fat "rewards" to get them through the day.[3] Feeling tired wipes out dietary restraint. Since Slow Dolphins are at risk for diabetes and heart disease, these food rewards might be setting you up for much bigger problems than sleep deprivation.

> ☀ Energize Hack: To get more sleep and choose healthier foods, eat on a fixed schedule during daylight hours.

SLEEP HACK: GUAVA LEAF TEA

Along with being antioxidant-rich, guava leaf tea has been found to boost immunity, lower cholesterol, prevent and treat diabetes[4]—and improve sleep quality. One reason Dolphins might wake up during the night is low blood sugar. That sets off a hormonal cascade that winds up turning on the cortisol tap. Cortisol is one of the circadian hormones that signal our master clock to start a new day, so it's not ideal for it to flow at 3:00 a.m. Guava leaf tea keeps blood sugar stable so that hormonal cascade never starts, and you stay asleep or calm down enough to fall back to sleep. We recommend the brand GuavaDNA, available online.

YOUR NEW KICK ASS ALL DAY, EVERY DAY PLANNER

Program these alarms on your phone for these key times:

7:00 a.m. Wake up; Stretch, plus water

10:00 a.m. Eating window opens

11:30 a.m. Bounce, plus water

3:00 p.m. Build, plus water

5:00 p.m. Bounce, round two, or exercise, plus water

6:00 p.m. Eating window closes

8:00 p.m. Shake, plus water

11:00 p.m. Power-Down Hour; turn off all screens

11:30 p.m. Balance (no water)

12:00 a.m. Bedtime

Week One

Begin on Monday. Any Monday. THIS Monday.

SLEEP ENERGY PRIORITY: CONSISTENT WAKE TIME AND BEDTIME

You have only two sleep energy tasks this week. *Make sure you get out of bed at 7:00 a.m. and get into it at midnight.* You will try to sleep only during those seven hours per day. If you stick with this plan for a week, you should see improvement in how long it takes you to fall asleep, and how long you stay down.[5]

> ☀ Energize Hack: By strictly limiting "in bed" hours, you train your body to seize the opportunity for sleep when it's presented to you. We call this *brain training*!

DAILY 5×5 PRIORITY: DOUBLE BOUNCE

To counteract fatigue and speed up metabolism, Slow Dolphins need to jump up and down, get their hearts pumping, and increase blood flow

264

and oxygen circulation to every cell in their body. We recommend double the Bouncing fun for your type. The morning Bounce helps break through brain fog, and the afternoon Bounce boosts energy expenditure (sugar burn) so that during your sixteen-hour fast, your body will access stored fat for energy more quickly. (See Chapter 4 for the exercises or go to MyEnergyQuiz.com for videos.)

- **Morning Bounce.** At 11:30 a.m., do two minutes of Jumping Jacks, followed by a one-minute rest. Then do another two minutes of Jumping Jacks.
- **Afternoon Bounce.** At 5:00 p.m., do two minutes of Jumps for Joy, followed by one minute of rest. Then do another two minutes of Jumps for Joy.
- **Optional.** Instead of Bouncing at 5:00 p.m., you can take a brisk walk or a bike ride for half an hour instead.

FASTING ENERGY PRIORITY: ESTABLISH AN INTERMITTENT FASTING RHYTHM

We recommend an eight-hour eating window from 10:00 a.m. to 6:00 p.m. All Slow chronotypes have only eight hours to eat, no more. In the first four hours of your sixteen hours of fasting, your body will use up all the carbs you ate that day (but if you move more, you'll burn them faster); then, for twelve hours, you'll be in fat-burning mode. Other benefits: By not eating within six hours of bedtime, your blood sugar rhythm will line up with your melatonin rhythm, and you'll fall asleep easily. Lock into this timing for circadian consistency!

MARC'S WEEK ONE ENERGY SCORE: 4

"I usually skip breakfast and eat super late, so the new eating schedule was a rocky adjustment. I felt queasy in the morning and very hungry late at night. But I do have to admit that after several days of eating this way, my appetite did shift. I'd been eating the same way for so long, I can see how I trained myself to get hungry at the wrong times. By eating at new times, I retrained myself. Instead of tempting myself, I avoided going for a drink after my shift and just went home instead. By the end of the week, I noticed that I had more money in my bank account, too, which was energizing as hell! The Jumping Jacks made me feel like an idiot. I'm sixty-one, not six. But I would go into the den, shut the door, and look like a fool on my own. The energizing effect was undeniable, but I told no one!"

Week Two

SLEEP ENERGY PRIORITY: LOVE YOUR BED

"Stimulus-control therapy" is a technique that gets you to stop thinking about your bed as if it were a torture device and to learn to associate it with sleep. Decades of scientific research have found the technique to be effective and reliable for the treatment of insomnia.[6] No more reading, watching TV, scrolling on FB, eating, talking, or working on that particular piece of furniture. Associating your bed with other activities perpetuates insomnia. Insomniacs tend to see it as a battlefield, where they go to suffer and struggle. To make peace with your bed, only use it for sleep and sex; cuddling is good, too.

DAILY 5×5 PRIORITY: STRETCH AND BALANCE

This week, add the morning and nighttime sessions to your day. Within a few minutes of rising at 7:00 a.m., take a quick trip to the bathroom to pee, and then get on the floor of your bedroom to Stretch for five minutes and ignite the pilot light of your day. At 11:30 p.m., Balance to focus the mind on anything but the threat of insomnia.

- **Stretch.** At 7:00 a.m., start with one minute of Child's Pose, then move into a minute of Cat-Cows. Do a minute of Sphinx, a minute of Dragonfly, and finish up with a minute of Cannonball.
- **Morning Bounce.** At 11:30 a.m. spend five minutes learning and mastering Burpees.

- **Afternoon Bounce.** Spend five minutes at 5:00 p.m. learning and mastering Ice Skater Jumps.
- **Balance.** At 11:30 p.m., do five minutes of Tree Pose. First stand on one foot for thirty seconds, then switch sides. Alternate feet until the time has passed.

THE MORE YOU MOVE, THE BETTER YOU GROOVE

Slow Dolphins need that extra motivational push, so consider adopting the mantra "The more you move, the better you groove." Although it might seem counterintuitive, the truth is, by expending energy by moving, you wind up with more of it. *Movement isn't a zero-sum game.* You don't "save" energy by being sedentary.

Think of it like a car battery. If you don't drive a car, the battery runs down. But every time you drive it, the battery is recharged. Slow Dolphins: *You are that car.* Take your body out for a spin. Not only will you recharge your battery; you'll *feel* better about yourself and your outlook on life.

FASTING ENERGY PRIORITY: SCHEDULING CAFFEINE

Drinking coffee and caffeinated tea, soda, and energy drinks is like walking a tightrope for Slow Dolphins. You might reach for those beverages in the morning to clear brain fog — which doesn't actually work; try fifteen minutes of sunlight instead. That won't affect your sleep later on. But if you have any caffeine after 3:00 p.m., it'll sabotage all of your efforts to teach your body to go down during those seven in-bed hours. We recommend that Dolphins have coffee or tea — just one 250 milliliters serving — with the midday meal. Skip it in the morning so you don't get jittery (as if you need more of that). Use an all-natural method to calm down, like meditation or sex (no instructions needed . . . we hope). And don't have any caffeine later in the afternoon or after dinner, ever. *Ever.*

> Energize Hack: Only have one cup of caffeinated beverage per day, with a pre–3:00 p.m. meal.

MARC'S WEEK TWO ENERGY SCORE: 4.8 (+.8 FROM THE PREVIOUS WEEK)

"Progress is measured in inches, not miles, and this week, I am definitely inching along. I feel like the eating schedule is locked in," he said. "I've been using the sleep techniques for two weeks and it's starting to work. It's physical and mental. When I look at my bed now, I long for sleep. And when I get into it, I drift off after an hour, which is a huge improvement for me. I can see how my old pattern of exhaustion and being sedentary only fed my insomnia, and I'm doing my best to change it. Last week, my wife pushed me to do the jumping exercise. But this week, when she forgot to remind me, I found myself pushing her to take five minutes from her work to do them with me. This is a 180-degree change. Now I'm the nag! That fact cracked me up all week, which gave me extra energy, too. And I thought I was too old for this shit."

Week Three

SLEEP ENERGY PRIORITY: POWER-DOWN HOUR

If your brain is running at a certain pace all day long, it doesn't just stop the second you want to fall asleep, especially for Dolphins. Your upside-down biology makes you feel most awake when it's time for sleep. It's not your fault, but you can do something about it.

The winding-down objective is to lower your blood pressure and body temperature to make a smoother transition from your energetic peak in the late evening into rest. For the other chronotypes, the process of vasodilation—blood leaving the body core and heading to the extremities—warms feet and hands but lowers the core body temperature, which coincides with the release of melatonin. If your feet are always cold in bed, it might mean your vasodilation circadian rhythm is screwed up, a common problem for insomniacs.[7]

It might seem counterintuitive to take a hot bath or shower before bed when you need to *lower* your core temperature. But when you get out of the tub, your body temperature drops. Forcing that drop is a biohack that flips on the sleep switch.[8] Another trick is to lower the room temperature to the National Sleep Foundation's recommendation of 60 to 67 degrees Fahrenheit, so crank the AC or open a window an hour before bedtime to chill the space.

Lower the temperature of your mind, too, by getting off all devices. Your fast, agile brain needs to disengage well before bedtime, or your thoughts will keep racing. Stop reading stimulating emails, status updates, tweets, articles, and blogs. The melatonin-suppressing blue light from an iPhone two inches from your face doesn't help either. Turn it all off. Don't worry. Any news that breaks before morning will still be there when you wake up.

Other recommendations for winding down: organize your desk, pack lunches for school, get into your pj's, pray, or try progressive muscle relaxation or meditation. Just unplug.

DAILY 5×5 PRIORITY: BUILD AND SHAKE

The last two sets to incorporate into your Daily Schedule this week are Build and Shake. You'll Build muscle mass at 3:00 p.m. to increase your power as you enter the late afternoon. At 8:00 p.m., it's all about Shaking off stress and stiffness so you can relax.

- **Stretch.** This week at 7:00-ish a.m., perfect your circuit of all five spine stretches: one minute of Child's Pose, one minute of Cat-Cows, one minute of Sphinx, one minute of Dragonfly, and one minute of Cannonball.
- **Morning Bounce.** At 11:30 a.m., go outside and Skip around the block for five minutes.
- **Build.** At 3:00 p.m., stop what you're doing and do Squats for one minute. Then do a minute of Crunches. Rest for a minute, and repeat.
- **Afternoon Bounce.** At 5:00 p.m., do five minutes of your jump of choice.

- **Shake.** When your energy goes up at 8:00 p.m., do one minute each of Neck Looseners, Arm Circles, Leg Swings, Crescent Bends, and Trunk Twists, with short rests in between.

- **Balance.** At 11:30 p.m., keep it grounded with five minutes of Tree Pose.

EMOTIONAL ENERGY PRIORITY: CHECK YOURSELF BEFORE YOU WRECK YOURSELF

A new, exciting Dutch study by researchers at the Netherlands Institute for Neuroscience looked at 2,224 insomniacs' life histories and came up with a classification system that divided them into five personality types.[9] As we've been saying all along, the more information you have about who you are, the better equipped you are to improve your mental and physical health. Our entire Energize system is based on being aligned with your DNA and working with what you've got to get the most out of your day and life. In that spirit, let's take a closer look at Dolphin personality types. Find yourself and use that information to figure out how your emotions and behaviors are draining your energy.

Type 1: Super anxious, super stressed. This type has classic insomnia symptoms of feeling most alert, active, and anxious at bedtime and, unfortunately, is prone to depression.

Emotional energy remedy: Movement. Movement is proven to affect mood in a positive way, even if you're severely sleep-deprived, and it'll help you get to sleep, too.

Type 2: Moderately stressed yet optimistic. Insomnia symptoms are most likely related to personal stressors and real problems. The saving grace for these highly emotional types is that they are "reward sensitive" and know how to make themselves happy.

Emotional energy remedy: Joy bringers. Do three things that bring joy every day until the difficult period passes.

Type 3: Moderately stressed and pessimistic. This type is driven to distraction and sleep deprivation due to high stress and anxiety. Unlike Type 2, though, Type 3 people are "reward insensitive," meaning they're pessimistic and unhappy and prone to depression.

Emotional energy remedy: Stop thinking. Insomnia is a thought-induced disorder. The more you think, the less you sleep. Redirect obsessive downward-spiral thoughts to spiral upward. Try the deep breathing exercise on page 253 to turn off the negativity and turn on serenity.

Type 4: Slightly stressed and highly reactive. This type loses sleep over life events, but their distress-based insomnia endures much longer than, say, a Bear's would if faced with the same hardship. They're "highly reactive" and will spin out over the slightest thing. Perfectionists fall into this category.

Emotional energy remedy: Stop fixing. Perfectionists tend to obsess over what they didn't do right and replay in their heads how they could have fixed a problem or what they'd do differently next time. First, acknowledge that you can't fix anything if you're exhausted. Then, make a note about your fixes or talk into your voice-memo app to download the thoughts from your brain so you can stop obsessing and get some rest.

Type 5: Slightly stressed and not too reactive. Trouble in this type's personal, financial, or professional lives can lead to sleepless nights, but they have "low reactivity," or less intense responses, than the previous types, as well as low motivation to change.

Emotional energy remedy: Do one thing. It doesn't matter what that one thing is. Just take positive control once, and the good energy you get from that will inspire you to do one *more* thing. One thing you can do is follow your Power Protocol to sleep better, move more, and feel more in control of your life. With that energy and confidence, you can change a bad situation, and a lifelong habit.

MARC'S WEEK THREE ENERGY SCORE: 6 (+1.2 FROM THE PREVIOUS WEEK)

"I'm a Type 3 insomniac, moderately stressed and reward insensitive. I do get down in the dumps, and a bad mood can spiral into mild depression. But I've found this week that by staying busy and always having something positive to do, I feel better about myself, and my anxiety and depression aren't showing themselves as often. Doing breathing exercises is never going to be my thing, but I've been taking long walks in the morning between my Stretch moves and when my eating window opens, and I've turned classically foggy time into useful time. I've gained hours, and energy."

Week Four

SLEEP ENERGY PRIORITY: SELF-COMPASSION

The thing about insomnia: Even when you follow the rules, bad nights will always happen, usually at the worst possible time. No one has perfect sleep, *nobody,* not even the Sleep Doctor himself. Even if you practice the techniques we've described, you'll occasionally find yourself lying awake, feeling like a failure at this basic human function. It's annoying and exhausting not to fall asleep, but it's not the end of the world.

You are not your insomnia. It's a side effect of being you, but it doesn't define you. Stop self-blame by practicing a self-compassion exercise. These two come from Kristin Neff, PhD, author of *The Mindful Self-Compassion Workbook: A Proven Way to Accept Yourself, Build Inner Strength, and Thrive.*

Exercise 1: How would you treat a friend? If your best friend were in the exact same boat as you and told you that she hated herself because of some body function that she was born with, what would you say to her? Would you tell her to blame herself and feel like a failure? Of course not! You'd encourage her to try some evidence-based strategies and do her best, and not to fault herself for her DNA. Biohack alert: The "friend perspective" forces your brain activity out of the amygdala's emotional center and into the prefrontal cortex, the logical, rational center. When you take the emotion out of the conversation, you can take control of your thoughts.

Exercise 2: Quiet your critical self-talk. According to Dr. Neff, the first step is to clue in to self-bashing when it happens. For you, it's probably while you're lying awake at night, ruminating and beating yourself up for not falling asleep. When you hear yourself doing it, reframe your thoughts toward compassion. For example, the usual self-talk might sound something like this: "What's wrong with you? Why can't you do this one thing that even babies can do effortlessly?" The friendly reframe: "You're an amazing wife/husband/mom/dad/sister/brother/daughter/son. You work hard and care deeply about others. You have so many gifts. Unfortunately, you do struggle to sleep. But that's just a tiny glitch in an otherwise lovable you."

This cognitive shift doesn't happen overnight. You might need to practice a nightly self-compassion exercise for weeks or months to internalize that a bad night isn't a personal failure. You may need therapy. But you will get there. You will sleep eventually. And every hour is an energizing gift. Appreciate the sleep you get. Tomorrow is another night.

DAILY 5×5 ENERGY PRIORITY: GO HARDER

This week, up the intensity.

- **Experiment.** Check the movement list in Chapter 4. Are there any you have been afraid of trying, like Dancer Pose or Gorilla Burpees? This is the week to face your fears and experiment with the movements that you've avoided.
- **Intensity.** Level up. If you have been coasting with Easy level, go Medium. If you've been doing Medium, do Hard. If you've been doing Hard level, up your amplitude. Jump higher, Skip faster, Squat deeper.
- **Duration.** Extend your Daily 5×5 to a Daily 6×5 or a Daily 5×7. Add a session or minutes if you are inspired.
- **Frequency.** If you've only been doing three sessions a day—some days are just like that—really push this week to hit the magic number of five.

FASTING ENERGY PRIORITY: SYSTEMS CHECK

After three weeks of intermittent fasting, you have already adapted to the change in your eating rhythm and have probably noticed that your clothes are a bit looser. This week push your window by fifteen minutes on either end to test your endurance. If you just *can't,* okay. But you won't know unless you try.

- **Open** your eating window at 10:15 a.m.
- **Close** it at 5:45 p.m.

MARC'S WEEK FOUR ENERGY SCORE: 7 (+3 SINCE WEEK ONE)

When he started his Power Protocol, Marc had several goals:

- **To stop feeling so tired during the day.** "I'm definitely a lot less tired. I don't really complain about it nearly as much. It was gradual, but when I look back at my energy level before I started, it's like I'm a different person now. I don't yawn constantly. I hardly doze on the couch. I'm not as grumpy."

- **Not to have to rely so heavily on coffee.** "It was a tactical change, to stop drinking coffee in the morning or at night. Now I have my mug at 2:00 p.m., and it's a treat I look forward to. To keep myself energized before and during a shift, I do Build and Shake instead of reaching for caffeine, and they work just as well."

- **To sleep better at night.** "Still a work in progress. But I have six straight hours a few nights in a row now, and that never happened before."

- **"My wife wants me to be more active, too, but that's her priority, not mine!"** "I'm definitely more active, so she's happy. And if she's happy, I am, too!"

Another life energized. Our work here is done.

TROUBLESHOOTING

Slow Dolphins, a few things to watch out for:

1. **Impatience.** Slow types have the most to gain from this program, but it'll take commitment. Your metabolism is like a cruise ship; you can spin the wheel, but it takes some time to change direction. We guarantee that you will lose weight and feel supercharged. Give it two weeks.

2. **Unrealistic expectations.** Insomniacs need to lose the illusion that, on any protocol, their sleep problems will disappear. It is not in a Dolphin's biology to get eight consecutive hours of deep sleep night after night. We can get you to six regularly; that's a win.

3. **Inconsistency.** Energizing comes into full force with *consistency*. Use your superpower of attention to detail by following the protocol as closely as possible, and you'll reap huge benefits very quickly.

4. **Frustration.** Some of the movements will feel awkward until you practice them enough. Ice Skater Jumps are hard to master. But who cares if you feel like a goof? Perfection is not the point. Silence the Olympic judges in your mind and give yourself top marks for effort.

Daily Schedule

The end of Week Four is the beginning of the rest of your high-energy life. Stick with it, and you'll always feel the power.

7:00 a.m. Wake up. 1 minute Child's Pose; 1 minute Cat-Cows; 1 minute Sphinx; 1 minute Dragonfly; 1 minute Cannonball; drink 500 milliliters of water

10:15 a.m. Eating window opens

11:30 a.m. 5 minutes Medium or Hard jumps of your choice; drink 250 milliliters of water

3:00 p.m. 1 minute Squats; 1 minute Crunches; 1 minute rest; 1 minute Wall Sits; 1 minute Dips; drink 250 milliliters of water

5:00 p.m. 5 minutes Medium or Hard jumps of your choice; drink 250 milliliters of water

5:45 p.m. Eating window closes

8:00 p.m. 1 minute Neck Looseners; a few seconds rest; 1 minute Arm Circles; a few seconds rest; 1 minute Leg Swings; a few seconds rest; 1 minute Crescent Bends; a few seconds rest; 1 minute Trunk Twists; drink 250 milliliters of water

11:00 p.m. Power-Down Hour; turn off all screens

11:30 p.m. 2 minutes Dancer; 2 minutes Eagle; 1 minute Figure Four

12:00 a.m. Bedtime

CHAPTER 15

The Other Power Profiles in Your Life

If you are curious about the energy ebbs and flows of your spouse, boss, parent, friend, siblings, adult children, and colleagues, all you have to do is ask them to take the body type and chronotype quizzes in Chapter 2 or at MyEnergyQuiz.com to lock in on their Power Profile. You might already have a sneaking suspicion of their energy type, but it's best to be sure. The quiz only takes a few minutes, and the insight they'll gain from it is invaluable—for them and for you in your future dealings with them.

In the case of someone very close, like a spouse or partner, the best idea would be to read the chapter about their Power Profile so you can sync up your schedules as much as possible. If you'd prefer to approach them when they're happy and alert, and avoid them when they're hangry or stressed, all you need are their SparkNotes to have energetically positive interactions.

If you have a relationship with a Medium Bear...

Here are a few things to keep in mind:

- ⬇ **Most sleepy:** Melatonin kicks in at 9:00 p.m.; sleepiness hits an hour or two later.
- ⬇ **Most hungry:** Bears wake up hungry.

⬆ **Most happy:** Peak serotonin hits in the late afternoon, from 3:00 p.m. to 6:00 p.m.

⬇ **Most stressed:** Early morning, from 7:00 a.m. to 9:00 a.m., is their touchy time.

⬆ **Most alert:** Peak focus and concentration happen from 10:00 a.m. to 12:00 p.m.

Medium Bears wake up grouchy and hungry, so it's best to avoid them until they have something to eat. Coffee will only make them jittery, so suggest water at breakfast.

Morning movement helps tamp down their bad mood at the start of the day. If they stick with their schedule and get some sunshine, by midmorning, they'll be back to their friendly, even-keeled selves.

Don't bother trying to connect with them after lunch. Their cortisol level crashes, and they'll be too tired to focus on what you're saying. Catch them around 3:00 p.m., when afternoon sleepiness shifts into their happy hour.

They thrive in group situations, and their mood will most likely rise if they're around people and the conversation is flowing. For a heart-to-heart, though, get them when they're relaxed at the end of the day and won't be too distracted.

Despite feeling tired at night, Medium Bears will stay on their phones and keep scrolling way past bedtime. But if they can unplug, they'll sleep better and won't be as touchy in the morning.

If you have a relationship with a Slow Bear...

Here are a few things to keep in mind:

⬇ **Most sleepy:** Melatonin kicks in at 9:00 p.m.; sleepiness hits an hour or two later.

⬇ **Most hungry:** Bears wake up hungry.

⬆ **Most happy:** Peak serotonin hits in the late afternoon, from 3:00 p.m. to 6:00 p.m.

⬇ **Most stressed:** Early morning, from 7:00 a.m. to 9:00 a.m., is their touchy time.

⬆ **Most alert:** Peak focus and concentration happen from 10:00 a.m. to 12:00 p.m.

Slow Bears, like Medium Bears, wake up grouchy and hungry, so it's best to avoid them until they have something to eat. Coffee will only make them jittery, so suggest water at breakfast.

When they're tired, they'd rather do just about anything other than jump up and down, so morning movements are a tough sell. But if they get in the habit — one move, one minute at a time — their morning mood will improve tremendously, and they'll act like their normal friendly, even-keeled selves.

Their cortisol level crashes after lunch, and they'll be sluggish and unfocused. Don't bother trying to engage until 3:00 p.m., when they get happy again and would love to chat with you.

All Bears thrive in group situations, and their mood will most likely rise if they're around people. But if they overeat, drink, and snack late into the night, their next-day energy will be rock-bottom. Encourage them to stick with an eating window of 10:00 a.m. to 6:00 p.m., and their all-day energy will perk way up.

For a heart-to-heart, get them when they're relaxed at the end of the day and aren't too distracted by whatever is on their screen.

Despite feeling tired at night, Slow Bears will stay on their phones and keep scrolling way past bedtime. But if they can unplug, they'll sleep better and won't be as touchy in the morning.

If you have a relationship with a Medium Wolf...

Here are a few things to keep in mind:

⬇ **Most sleepy:** Melatonin kicks in at 11:00 p.m. to 12:00 a.m.; sleepiness hits an hour later.

↓ **Most hungry:** Wolves' appetites kick in around 12:00 p.m. or later.

↑ **Most happy:** Peak serotonin hits in the evening, from 7:00 p.m. to 10:00 p.m. or later.

↓ **Most stressed:** Morning is their touchy time.

↑ **Most alert:** Peak focus and concentration happen from 2:00 p.m. to 6:00 p.m.

Medium Wolves can barely see straight in the early morning, so the best thing for their mood and energy is sunlight and movement, ideally at the same time. Hydration will also wake up their digestive systems if they're not hungry enough to eat.

Their energy picks up at noon, and that's when their day really begins. By afternoon, their sharp minds are crackling; the end of the workday is their time to shine in small meetings. They're better one-on-one or in groups of three or four.

Wolves are highly social (when they want to be) but need plenty of alone time. If the door is closed, knock gently and ask if you can come in. If the answer is "no," don't take it personally. It's not about you. When they say "yes," they will welcome you with affection, curiosity, and focus, as long as it's later in the day.

Getting Medium Wolves to scale back on coffee will be a challenge, but if you can convince them to replace caffeinated beverages with water, they'll get a twofer of energy due to increased hydration (Wolves forget to drink) and more sleep (which they always need).

Their rebellious nature might have them resist movement. If they can look at movement as a new addiction, they'll become obsessed with it and try to get everyone they know involved.

If you have a relationship with a Slow Wolf...

Here are a few things to keep in mind:

⬇ **Most sleepy:** Melatonin kicks in at 11:00 p.m. to 12:00 a.m.; sleepiness hits an hour later.

⬇ **Most hungry:** Wolves' appetites kick in around 12:00 p.m. or later.

⬆ **Most happy:** Peak serotonin hits in the evening, from 7:00 p.m. to 10:00 p.m. or later.

⬇ **Most stressed:** Morning is their touchy time.

⬆ **Most alert:** Peak focus and concentration happen from 2:00 p.m. to 6:00 p.m.

Slow Wolves hate the mornings, especially if they have to start jumping up and down first thing. But if they can push past that resistance, their energy will go way up and they'll be somewhat functional before noon.

Slow Wolves are at a high risk for anxiety and depression. If they stick with the five daily movements schedule, they'll sleep better, which can reduce the risk of mood disorders. Wolves aren't known for being conscientious, but they do love to try new things. To encourage them, focus on the novelty of a new lifestyle. Once they get rolling with the program and the pounds come off, they'll feel better about themselves.

Their minds are sharpest at the end of the workday. If you approach them with questions and concerns then, they'll have brilliant answers for you. They're most creative and philosophical at night. Ask them for personal advice then, and you'll get plenty of insight.

Wolves are social (when they want to be) but need plenty of alone time. When they're into being with people, they're the life of the party. Otherwise, back away from them slowly to give them plenty of space.

Substance abuse, whether alcohol, recreational drugs, sugar, or caffeine, might be an issue. To support the Slow Wolf in your life, take care of the practical things, like getting that crap out of the house. And give them plenty of love and encouragement. Happy hormones like oxytocin and dopamine can be a natural, ready replacement for drugs.

If you have a relationship with a Fast or Medium Lion...

Here are a few things to keep in mind:

- ⬇ **Most sleepy:** Melatonin kicks in at 7:00 p.m. to 9:00 p.m.; sleepiness hits an hour later.
- ⬇ **Most hungry:** Lions wake up hungry.
- ⬆ **Most happy:** Their cortisol drops (and serotonin rises) from 1:00 p.m. to 5:00 p.m., so they have relatively calm afternoons.
- ⬇ **Most stressed:** With melatonin rising from 7:00 p.m. to 9:00 p.m., they fight to stay awake.
- ⬆ **Most alert:** Peak focus and concentration happen from 6:00 a.m. to 1:00 p.m.

Morning workouts are almost a requirement for Fast and Medium Lions to burn off the rocket fuel they wake up with. (Every-other-day workouts allow their muscles a chance to rest and recover.) Introverts, Lions prefer solo fitness, so don't suggest joining them on a run; it'll get awkward.

Their rocket fuel keeps them going all morning long. During their peak alertness time, they *need* to get things done. If you have important business to discuss, you will get a decisive thumbs-up or down from them before noon. Lions who power nap can stretch their alert time into the early afternoon.

Lions are on a *mission*. So if you present them with a problem, they'll rush to fix it. Sitting around and validating your feelings will make them antsy and uncomfortable. It's not that they don't care. They'd just rather be active than passive.

Hydration is a problem for all Fast and Medium types, so whenever you need to have a conversation with a Fast or Medium Lion, bring a non-caffeinated beverage with you. At the end of the day, when they start to fade, a big drink of water will perk them right up.

If you're after fun social time with a Fast or Medium Lion, schedule it for the late afternoon or early evening, their twilight chill hours. But don't push it until nighttime or you'll lose your audience when they can't keep their eyes open.

If you have a relationship with a Slow Lion...

Here are a few things to keep in mind:

- ⬇ **Most sleepy:** Melatonin kicks in at 7:00 p.m. to 9:00 p.m.; sleepiness hits an hour later.
- ⬇ **Most hungry:** Lions wake up hungry.
- ⬆ **Most happy:** Their cortisol drops (and serotonin rises) from 1:00 p.m. to 5:00 p.m., so they have relatively calm afternoons.
- ⬇ **Most stressed:** With melatonin rising from 7:00 p.m. to 9:00 p.m., they fight to stay awake.
- ⬆ **Most alert:** Peak focus and concentration happen from 6:00 a.m. to 1:00 p.m.

Morning movement is a requirement for Slow Lions to burn off the rocket fuel they wake up with and to stave off hunger until their eating window opens at 10:00 a.m. Until they have their first meal, though, they might not be too keen to have personal chats.

Their rocket fuel keeps them going all morning long. During their peak alertness time, they *need* to get things done. If you have important business to discuss, you will get a decisive thumbs-up or down from them before noon. Lions who power nap can stretch their alert time into the early afternoon.

Lions are on a *mission*. So if you present them with a problem, they'll rush to fix it. Sitting around and validating your feelings will make them antsy and uncomfortable. It's not that they don't care. They'd just rather be active than passive.

Slow Lions' minds are always climbing mental mountains, but their bodies might be too sedentary. A great way to engage a Slow Lion is to get them to take a walk around the block or run an errand with you in the early afternoon. They need the movement. You'll get them away from their desk, and they'll be most receptive to hear your ideas then.

If you're after fun social time with a Slow Lion, schedule it for the late afternoon or early evening, their twilight chill hours. But don't push it until nighttime, or you'll lose your audience when they can't keep their eyes open.

If you have a relationship with a Fast or Medium Dolphin...

Here are a few things to keep in mind:

- ↓ **Most sleepy:** They are crushed with fatigue in the early morning, from 7:00 a.m. to 9:00 a.m.
- ↓ **Most hungry:** They get super hungry and should have some protein at 9:00 a.m.
- ↑ **Most happy:** They have creative, chill, daydreamy time from 10:00 a.m. to 12:00 p.m.
- ↓ **Most stressed:** Their cortisol rises and stays high starting at 10:00 p.m.
- ↑ **Most alert:** Peak focus and concentration happen from 4:00 p.m. to 6:00 p.m.

Dolphins' genetically high stress levels respond well to intense movement early in the day, but they still need to complete their five daily movement sessions to get a huge sense of accomplishment. Morning exercise really sets the pace for lower anxiety.

Be supportive of Dolphins and the love will come back to you. They are some of the most loyal friends in the world.

Working or hanging in groups can be stressful for them since they do have a bit of social anxiety. But the more they're drawn out of their shell, the more practice they'll get at interacting. Encourage them to work and play in groups, but don't push.

Hydration is important for Fast and Medium Dolphins since they are always fidgeting and moving. Buy them a water bottle with measurements on the side.

Dolphins can get touchy if they forget to eat, so it's a good idea to remind them to have a meal.

The best time of day to discuss important topics with Fast and Medium Dolphins is after they've eaten dinner at 7:00 p.m. and before their nightly cortisol kicks in at 10:00 p.m. That's their interpersonal sweet spot.

If you have a relationship with a Slow Dolphin...

Here are a few things to keep in mind:

- ↓ **Most sleepy:** They are crushed with fatigue in the early morning, from 7:00 a.m. to 9:00 a.m.
- ↓ **Most hungry:** They get super hungry and should have some protein at 10:00 a.m.
- ↑ **Most happy:** They have creative, chill, daydreamy time from 10:00 a.m. to 12:00 p.m.
- ↓ **Most stressed:** Their cortisol rises and stays high starting at 10:00 p.m.
- ↑ **Most alert:** Peak focus and concentration happen from 4:00 p.m. to 6:00 p.m.

Whenever Dolphins feel overwhelmed, encourage them to jump up and down to get endorphins flowing and to reset their thoughts. Morning movement really sets the pace for a low-anxiety day.

Since Dolphins are exhausted in the morning, it's not a great time to try to talk to them about anything important. Or anything at all. Just back off until they're more awake around 10:00 a.m. If you need something before then, send the request by email or text.

They'll feel a huge sense of accomplishment if they stick with their eating window and five daily movements schedule. Due to their chronic exhaustion, Slow Dolphins can be too sedentary. Support them by doing the movements with them, and the love will come back to you. Dolphins are some of the most loyal friends in the world.

Working or hanging in groups can be stressful for them since they do have a bit of social anxiety. But the more they're drawn out of their shell, the more practice they'll get at interacting. Encourage them to work and play in groups, but don't push.

The best time of day to discuss important topics with Slow Dolphins is after they've eaten dinner by 6:00 p.m. and before their nightly cortisol kicks in at 10:00 p.m. That four-hour window is their interpersonal sweet spot.

PART III

ENERGY FOR LIFE

CHAPTER 16

Year-Round Energy

We've provided ideal protocols for maximum energy for a normal day, but as we all know, not every day is the same throughout the course of the year. We all have energetic cycles. To get through them, be accepting and gentle with yourself.

> ⚡ Energize Tip: During any period of change from your normal routine, pay *even more* attention to your sleeping, moving, and eating rhythms to maintain the energy level you've grown accustomed to. Even exciting, positive experiences take a lot out of your body battery.

WINTER BLUES

Seasonal affective disorder (SAD) is a mood disorder and an unforgiving energy drain. It strikes at the end of fall and the beginning of winter and usually lifts in springtime. If you find your energy shifting to "low" and "down" when the days get shorter, you might have SAD. Symptoms include:

- Depression
- Loss of interest in what used to bring you joy
- Sleep issues
- Overeating
- Drop in sexual desire
- Low energy
- Feeling easily aggravated

- Hopelessness
- Difficulty concentrating
- Low motivation to be around people

The dark months mean less exposure to sunlight. And since sunlight is a major zeitgeber, winter is disruptive to our circadian rhythms and hormonal balance. Too much melatonin makes us feel tired; too little serotonin makes us feel lousy. For an Australian study, researchers tracked their 101 participants' level of the happy hormone serotonin throughout the year and found it to be lowest in winter.[1] You're not imagining it. The change in season *does* accompany a downgrade in mood. And when mood is low, people tend to withdraw and cut themselves off from powerful mood and energy boosters like social interaction, sex, shared laughter, and music.

Energy gains can be found during the late fall and winter months if you know where to look. Here are some recommendations to beat SAD for all energy types, but to be clear, you should connect with a mental health care provider to approve any of the following recommendations:

- **Take vitamin D supplements,** if you're not already; 600 IU daily should do it.
- **Grab any sunlight you can get.** It might mean putting on your parka and boots, which is a universally acknowledged huge pain in the ass. But you need the sun to stimulate your optic nerve and travel to your master-clock center to keep your circadian rhythms in sync. Shoot for five minutes of outdoor time after your Stretch and fifteen minutes after (or during) Bounce.
- **Don't skip your Daily 5×5.** A wealth of studies have found that regular exercise—even five daily movement sessions!—is an effective and accessible treatment for SAD.[2] Movement boosts mood, as you now know by heart.

THE HOLIDAY SEASON

The most wonderful time of the year—the holiday season—rides in on a snowbell-strewn sleigh with two major energy drains: (1) travel and (2) stress, family related and otherwise.

↓ **Beware Jet Lag**

If you cross time zones, you'll experience symptoms of jet lag—fatigue, fogginess, short-term memory deficiency, insomnia—and they won't magically disappear like Christmas cookies or your bank balance after holiday shopping. According to a University of California, Berkeley study, researchers subjected poor hamsters to the circadian disruption of jet lag and then compared them to a control group.[3] The first group's memory and learning ability was far inferior to that of the nontweaked hamsters, and their mental-energy drain lasted for a month!

Generally, it takes one day to recover from crossing one time zone. So if you cross six, you might be impaired for six days. And it's always worse when traveling east.

Some tips for regaining energy ASAP when traveling:

- **Avoid alcohol and caffeine.** Both are dehydrating and sleep disruptive.
- **Move!** A brisk walk every morning at your destination will help reset your rhythms, as will adjusting your movement schedule to your new time zone.
- **Don't eat the cookies.** Leave them for Santa, seriously. A sugar rush and the resulting insulin spike will have you dozing on the couch, a bad idea when you're already dealing with jet-lag sleep adjustments.

OUR JET-LAG ANTIDOTE

We both use Timeshifter.com when we travel. You plug your chronotype, where you are, where you are going, and the time of your flight into this app. Then it gives you a personalized schedule for when to get light exposure, consume caffeine, nap, and take melatonin, so you arrive on the schedule of your destination!

The source of holiday-related stress might be the pressure to be merry, the added burden of making plans and buying gifts, and, of course, the potential for toxic interactions with family members. Dolphins beware: Interpersonal stress causes greater severity of insomnia and even worse rumination than usual.[4] To manage holiday stress, **prioritize sleep** and, regardless of your food options, eat everything (slowly) in moderation and start with a small plate. You can always go back. Everyone loves healthy leftovers!

> ⚡ Energize Tip: Wolves, Lions, and Dolphins who all need adequate "alone time": There is no reason to feel shame or guilt about needing space. By going off on your own for a few hours a day, you can recharge and be fully present at family gatherings. Or stay at a hotel or nearby Airbnb, so you can get a little space.

HOLIDAY ENERGY DRAIN: EMOTIONAL VAMPIRES

Emotional Vampires are the toxic people who suck your energy dry with constant complaining, blaming, shaming, making demands, ultimately crossing boundaries, and asking for help and favors, reassurance, and validation. *The easiest way to recognize them is by paying attention to how they make you feel.* Don't try to keep up with their stream of consciousness. Sometimes you just need to be the listener, not the fixer. Clue in to your body sensations instead. Do you start to feel a heavy weight on your shoulders? Do your muscles stiffen from tension? Do you suddenly need to lie down?

When an Emotional Vampire tries to latch on to you:

- **Keep it short.** Sit in a state of empathy and don't carry on the conversation too long.
- **Slip away.** When the time is right, gently remove yourself from the situation and try to find a safe space to reset. Leave on a compassionate note.

At family gatherings, set an Energy Agenda to hang out with those who lift you up with laughter, joy, affection, optimism, connection, and interest in your life. Seek human energy rockets—and be one! If you do, you'll have a great time at the party, guaranteed.

VACATIONS

Vacations can be a time to recharge your battery and get a long-deserved break from the stresses of work and demands of life. But the one thing you do *not* want to do is take a break from your Energize rhythm.

Oh, we get it. Throwing out the "rules" and relaxing is exactly what we look forward to on our vacations as well. But the unfortunate fact is that while oversleeping, overeating, overdrinking, and underexercising might feel like indulgent luxuries while you're in "away" mode, those energy drains are a setup for exhaustion when you return to normal life. It's why people say cheekily, "I need a vacation from my vacation."

We are not wet-blanket killjoys who want to take away your piña coladas. That daybed on the beach has your name on it. Enjoy! Nap! Be present and disconnected at the same time! Do as little as possible... while sticking with the broad parameters of your Power Protocol.

Remember the basic math: As long as your energy gains are greater than your energy drains, you'll be able to *go* and *do* while on vacation and when you get home.

On your trip, shoot for a balance of indulgence and diligence and make strategic choices to mitigate drains and

> ⚡Energize Tip: If possible, schedule a day for reentry so you don't scramble to come back to reality.

boost gains. If you do that, you'll show up for work on the Monday after vacation with a radiant tan and an energetic glow.

Step 1: Prioritize. What indulgence is most important to you: eating outside your eating window, sleeping late / staying up late, or being sedentary? If you're on a trip to Paris, it's probably eating and drinking! If you're going camping, it might be sitting on a log and not moving (except to cast a line) all day. If you're staying in a beach resort and don't have kids, it might be sleeping until noon and staying up way past midnight. Prioritize which indulgence is most important to you.

> ⚡ Energize Tip: Choose one category to indulge and be extra vigilant with the other two.

Step 2: Modify, aka indulge within reason. So if you choose to indulge in the eating category, modify your schedule instead of completely abandoning it.

To modify your eating rhythm, Slow types, open your eight-hour eating window to ten hours or even twelve. Medium types, open your ten-hour window to twelve.

Twelve hours is the maximum eating window, or your vacation will be ruined with indigestion, gas, and bloating. If you wind up drinking until midnight, skip breakfast. We know that all-you-can-eat breakfast buffet looks tempting. But it'll be there tomorrow.

Regarding sleep modification, if you want to have lazy mornings, set the alarm no later than one hour after your usual wake time, and plan a luxurious afternoon nap into your day. Get in bed no later than one hour past your usual bedtime. For quality sleep, stop drinking three hours before bed.

To modify a movement schedule, commit to three of five daily sessions on top of whatever physical activities you have planned. We recommend always doing morning Stretch, midday Bounce, and pre-bed Balance (anchor yourself with a chair if you've had a few adult bevs, *please*). That way, even if you are lying on a pool lounger all day, you will get some movement and up your likelihood of getting quality sleep.

Step 3: Forgive. Okay, so maybe you completely abandoned your protocol for the entire vacation, and now you feel guilty as well as exhausted. Do not beat yourself up for one second. Just assume your Energy Posture (sit up straight!), do some calming diaphragmatic breathing, and say, "I'm psyched to *energize!*" Then get right back into the program that you already know will restore your power and make you feel awake, alert, and alive. It only takes a few days to get back on track, as long as you don't drain energy on a useless emotion like guilt. Let. It. Go. And move on. Use that positive inner voice we always talk about. Self-shame is a one-way ticket backward.

A SEASON OF ANXIETY, DEPRESSION, OR GRIEF

We all go through seasons of loss or upheaval that are exhausting, to say the least. Upsetting emotional experiences like anxiety, depression, and grief make you want to crawl into bed and pull the covers over your head.

Trying times can temporarily increase your need for sleep. If stress is chronic, it can create a chronic sleep debt. (And it's not just negative or unwelcome life events that can drive up the need for sleep: Any big life changes, including happy ones like a new job, home, marriage, or baby, can demand more sleep, too.) Grief, for example, has been found to cause prolonged sleep disruption. A recent New York University School of Medicine study found that baseline sleep disturbances occurred in 91 percent of their 395 patients with intense grief.[5]

Depression and sleep are a chicken-egg situation. We're not sure which came first, but we do know they go together. Ninety percent of patients with depression report poor sleep.[6] Anxiety causes insomnia, as ruminators of every type know from personal experience. If you have insomnia, you're up to seventeen times more likely to struggle with anxiety and depression.[7]

You know the cycles of energy by now. If you're not sleeping enough, you're more likely to crave foods that torpedo energy, which leads to not moving, which makes it harder to sleep, which increases anxiety . . . and around we go. Your Power Protocol is never more essential to your health and well-being (and energy) than when you are going through something exhausting.

Wolves of all metabolic speeds are particularly vulnerable to "off" seasons, so take note: If you are going through a rough patch, muster whatever energy you have to be proactive about sleep, if nothing else, and we'll deal with exercise later.

- **Rely on the comfort of a set routine.** Your Power Protocol is one thing you don't have to worry about. It's there for you to give your life structure and to make you feel better.
- **Take something.** If you're locked in a cycle of insomnia or sleep disruption because of grief, depression, or anxiety, this is the time to consult with your doctor about using prescription sleep aids. The death of a loved one, financial turmoil, a relationship split . . . the drugs exist, and they can take some of the pressure away. If using sleep meds is not a good idea for any reason (like a prior addiction), try non-habit-forming options like the Sleep Doctor PM Night Time Formula or melatonin supplements (at the right dose!).

With treatment and time, adequate rest, exercise, and nutritious food, you'll get through "off" seasons and reclaim your energy to fight another day. With a fully charged body battery and a full heart, you can overcome anything.

↑ Emotional Energy Gain

Get help. A licensed therapist can treat your mood and/or sleep disorders. It's up to you to seek one out. Go to your physician to discuss what's going on. They can refer you to a specialist if needed.

Monthly Energy Rhythm

Monthly cycles take their toll and leave many women exhausted from hormonal ebbs and flows and the pain of cramps. Yogis call it "moontime."

In any study of chronotypes, if there's a negative finding, it's usually associated with Wolves. Menstrual woes are no exception! According to a Finnish study of 2,672 women, evening types have the longest duration of bleeding compared to intermediate and morning types. Lions, of course, have the shortest periods.[8]

Wolves: There's just no rest for the rebellious.

If you can accurately predict the timing of your periods, you can better prepare for them. If your master-clock rhythm is reinforced by following your Power Protocol, your periods are more likely to be regular.[9]

> ✸ Super Energize Hack: Use a menstrual cycle timing app to help you out, then cross-reference it with your Power Protocol.

THE ENERGETIC FLOW OF YOUR MONTHLY FLOW

The week before, aka the late luteal phase: Some women might have restless sleep during this phase. A recent study tracked the reproductive hormone levels in eighteen- to twenty-eight-year-old women over two complete menstrual cycles.[10] Leading up to their periods, estrogen and progesterone dropped, as did the participants' sleep efficiency. It took longer to fall asleep, and they woke up more often during the night. Possible explanations are premenstrual women's increased body temperature[11] and, for some, shortened REM sleep.[12] What's more, the women were hungrier during that phase, too, due to the drop in serotonin and appetite-controlling estrogen.

Best movement sessions this week: Stretch and Balance. Try to avoid strenuous exercise and stick to low-impact moves in your Power Protocol. Bloated bodies may be sensitive to bouncing up and down.

The week of, or the menstrual phase: Energy is lower when you get your period, in part due to sleep disruption. But after a day or two, estrogen and serotonin will rise again. By the end of your period, you'll feel sleepy again.

Best movement session this week: Bounce. Every woman experiences her cycle differently. Stay connected to your body. Get what *you* need from each day in regard to exercise. If you are up for it, Bounce will help you to sweat out water retention and make you feel more "together."

The week after, aka the early follicular phase: This is your most energetic week of the month, a result of body temperature regulation and peak estrogen and serotonin flow.

Best movement sessions this week: Build and Shake. Rising estrogen means more stamina, higher tolerance for pain, and an opportunity to build muscle. Go long and hard with strength-building moves. There is evidence that joint flexibility is highest during hormonal fluctuations.[13] This week, while your hormones are stable, you'll be stiff, so be sure to Shake daily to loosen up.

The week after *that,* or the ovulation phase: To cue ovulation, estrogen drops suddenly, which can cause temporary insomnia for some women. But then progesterone, a sedating hormone, rises and brings estrogen along with it for your best sleep of the month.

Best movement session this week: Bounce. With all the great sleep you're getting now, your body is rested and ready for movin' and groovin', so double Bounce sessions while you have energy to burn.

TAKEAWAYS: YEAR-ROUND ENERGY

• Energy is not static. It ebbs and flows daily, weekly, monthly, and seasonally. But in each ebb, you can counteract a draining time with energizing strategies.

• Draining seasons are wintertime, the holiday season, and "off" seasons of emotional upheaval. During those times, take extra care to stick with a modified version of your Power Protocol so that you'll be strong and healthy when that draining season passes.

- Vacations are a great time to catch up on sleep and stockpile emotional energy gains. But if you can stick to a modified version of your Power Protocol, you won't need a vacation from your vacation to deal with the exhaustion of returning to a normal routine.

- Women: Your menstrual cycle brings intense hormonal fluctuations that can cause roller-coaster emotions. Anticipating how energy changes week to week throughout the month can help you compensate for drains and take advantage of gains.

CHAPTER 17

Energy and Your Health

Each protocol is designed for optimal sleeping, fueling, and moving and is a plan for overall health and well-being to *prevent* illnesses and injury. Ideally, you will never get sick or hurt or suffer a minute's pain for one tiny moment. But if you do, you can use Energize practices to speed your recovery.

ILLNESS

Illnesses like colds, flus, and viruses (we shall not speak its name) are always energy draining. It's not so easy to stick with a schedule—or even lift your head—when you feel awful. But even when you're flat on the bed and barely have enough energy to watch *The Mandalorian,* standard energizing practices—the ones you've already been doing—will boost your immune system to speed healing.

- **Eat plenty of fruits and vegetables.**
- **Move regularly.**
- **Stay hydrated.**
- **Get adequate, high-quality rest.**

The last two items on that list—hydration and rest—are the most important Energize strategies while you're actively fighting off an illness.

Dehydration is linked to gastrointestinal and kidney problems,

impaired cognitive and physical performance, heart disease, skin conditions, and headaches.[1] It also slows down the transportation of nutrients,
oxygen, T cells, which fight illness, and B cells, which contain antibodies
to our organ systems. It's called a "blood*stream*" for a reason, since it's
largely composed of water. When adequately hydrated, our bloodstream
flows like an engorged river, sending those immunity cells and nutrients
where they are needed to fight disease; a well-hydrated lymphatic system
flushes out cellular waste, aka toxins. A dehydrated bloodstream and
lymphatic system are like a dried-out riverbed. Nutrients can't get into
our organs; cellular waste products
can't get out. Drinking the requisite four
or five 250 milliliters glasses of water
per day during an illness not only gives
you energy; it helps clear your body of
the germs that are making you feel like
crap. Consider drinking a CBD water
like Akeso for maximum hydration and
inflammation-crushing benefits.

> Energize Tip: You might not
> want to chug water, herbal tea, or
> clear broths when you have a cold
> or the flu, but if you can take sips
> every fifteen minutes, you'll feel
> better sooner. Personally we love
> chicken soup, aka Mom's penicillin!

Rest is the immune system's war room. While you're sleeping, your
body assesses the threat of whatever germs are on the attack, then comes
up with a strategy to fight them off. Without enough sleep, your body
can't organize its defenses or devise a plan. Sleep also boosts production
of T cells, those white blood cells that are the body's foot soldiers, the
front line of defense. In one recent German study, researchers measured
the T cell activation levels of participants who got a full night's sleep and
of those who did not.[2] The exhausted group's immunity response was
lower than that of the well-rested group.

A recent University of California, San Francisco study highlighted
what we all know to be true: Short sleep
makes us more susceptible to getting
sick. Researchers tracked the study's 164
participants' sleep patterns for seven
days.[3] Then they were quarantined in a
hotel and intentionally infected with a

> Energize Tip: Following the
> sleep recommendations in your
> Power Protocol will help prevent
> *and* cure illnesses.

rhinovirus (aka the common cold) via nasal drops. Those who'd tracked less than seven hours of sleep the week before were 4.2 times more likely to catch the cold than the well-rested group.

INJURY

An injury due to overdoing it at the gym or on the field, or due to an accident, is exhausting because (1) your body needs to direct the force of its energy toward healing what's broken, (2) being sidelined is frustrating— no one wants to feel like they can't do something, and (3) pain is an energy drain like no other.

Some chronobiologic ounces of prevention:

Wolves, you are more likely get in a car accident in the morning. For a recent study, researchers tested morning and evening types on their driving performance at 8:00 a.m. and 8:00 p.m. and found that evening types were more prone to driving errors during the early test sessions, their nonpeak time of day.[4] Morning types were conscientious drivers during off-peak and peak test times.

Wolves, if you get behind the wheel in the morning, you're at risk for an accident. Muster extra vigilance by limiting distractions and focusing on the road. Be sure to get ten minutes of sunlight before you drive!

Dolphins, you don't get off so easy. Chronic lack of sleep is a huge risk factor for car accidents. Per the National Highway Traffic Safety Administration, drowsy driving caused 795 deaths and 91,000 crashes in 2017. And if you play sports or work out when you're exhausted, you're more likely to suffer a sports injury.[5]

Lions are susceptible to overtraining injuries, especially at their peak energy time, in the early morning. You might be raring to go, mentally pumped, and saying Lionesque things like "Let's *do* this," but if your body isn't warmed up and stretched out, you might pull something (which is one of the reasons we have you Stretching first thing, before you lace up).

Bears, the weekend warriors, are at their greatest risk for injury when they go hard sporadically. Compressing physical exertion into small bursts and pushing yourself too hard when you don't have the conditioning (sorry) can lead to pulls and strains. Pace yourselves, Bears! Either work out more consistently, or don't push too hard when you do exercise.

STACEY SAYS...

Just a few pointers about healing from a sports injury.

Be patient! I know you want to get right back out there, but you'll only make it worse if you start exercising again too soon. I've had clients who've insisted that they were totally healed from a leg injury, and by the end of class, their knees were the size of grapefruits. Give yourself six weeks to recover and get the all clear from a doctor or certified physical therapist before you get back on the bike or on the road. And take it easy for the first two weeks. Gradually up your amplitude until you're back to being your badass self. Now more than ever is the time to focus on your diet, and your *sleep*, especially when you are sidelined, okay?

Take it seriously. Rate the level of pain from one to ten; if you're at a six or above, you are *injured*. I've seen clients deny that they had an injury at all, even though they were limping. Once I convinced one to get an X-ray; it turned out she had a hairline fracture in her foot. Don't act like nothing's wrong if something is definitely hurting. It's not about being able to run today. It's about being able to walk for the rest of your life.

Train the part that's not hurt. You don't have to sacrifice the benefits of movement when injured if it's safe to train the non-injured parts. If you broke your leg, you can still lift hand weights. If you broke your arm or fractured your wrist, you can still take long walks. As long as the injured part isn't stressed, you can and should stick with your five daily movement sessions.

OH, THE PAIN!

When in physical pain, using the Energize method for relief might seem counterintuitive, but it has been proven to work. A well-studied phenomenon called "exercise-induced analgesia" works by promoting pathways for the neurotransmitters endorphins and serotonin in the brain so that when you move, you get a hormonal high that reduces pain.[6] An extensive review of 381 studies, encompassing 37,413 participants, compared exercise vs. no exercise interventions to relieve pain caused by conditions like rheumatoid arthritis, osteoarthritis, fibromyalgia, low back pain, neck disorders, and spinal cord injury.[7] Researchers concluded that exercise reduces pain severity, boosts physical functioning, and provides psychological benefits, which all contribute to an improved quality of life. This is a huge and complex subject, with specific recommendations depending on the cause of pain and fitness level. *Anyone who has chronic pain must talk to their doctor before starting any exercise regimen to reduce pain.*

TAKEAWAYS: ENERGY AND YOUR HEALTH

- Following the sleep recommendation in your Power Protocol will help you prevent *and* cure illnesses.
- All Power Profiles are more likely to get into a car accident at their non-alert time of day. Wolves are at risk in the morning, Lions in the evening, Bears in the early afternoon, and Dolphins in the morning.
- Injuries are more likely when the body isn't warmed up, so even Lions are more susceptible first thing in the morning.
- If you do have an injury, no matter how frustrating it might be to be sidelined, energize by resting and healing until you're cleared by a medical professional.
- Physical activity has been proven to help manage pain. As long as your doctor approves, use the Daily 5×5 to help relieve pain.

CHAPTER 18

A Word Before You Go, Go, Go!

You now have all you need to live energized. Just follow your maintenance schedule (located at the end of your Power Protocol chapter). If you fall off, it's okay. Pick it back up and pat yourself on the back for taking care of yourself. Some final thoughts and key takeaways from each of us:

STACEY SAYS...

It's not always easy to realize things about yourself when circumstances throw you out of your protocol. Your body may be telling you one thing, but your reality is telling you another. I can recall a specific time in my life when I was diagnosed by a doctor as being sleep-deprived, but I didn't listen. My days were packed with physical work to the point of exhaustion, and my nights were filled with social and family obligations that I just couldn't keep up with. I imagine that this is a state that many people reading this book can relate to. You have long days, long nights with social commitments, homework assists, or partners who may be on an opposite schedule both physically and mentally. One of the most challenging things in life to overcome, I have found, is the ability to mesh *your* patterns with those of other people. Whether it's family, friends, or work associates, *your pattern* has to intertwine with those around you, and when it doesn't, it leads to a feeling of disruption.

This book is highly prescriptive. The chronotypes are meant for you to identify the methods that work for *you*. The key to making your new protocol work for you is how you integrate it into your existing life! Whether that means shutting down your day earlier without backlash or pushing your morning start times a little further out, make sure you identify which parts of your life will be the most accommodating to your new changes, as some of them may be easier to adjust than others. I have found that honesty and a gentle approach will be most impactful. Just remember that most people are uncomfortable with change, and although you may have the best intentions, change is not always well received by others.

I encourage you to stay strong with your commitment to yourself, and give your new outlook and protocol some time to actually kick in. Remember that things don't happen overnight; they take time. Every day that you stay connected to your chronorhythm is another step forward into living a life that is specifically suited to your human machine. Imagine waking up every day with more energy, more zest, and more mojo! Sleep, exercise, and a healthy diet are the quintessential keys to life! In that order! Remember that when someone says, "I don't need sleep," they just aren't aware of the *insane* benefits of it, and those who choose not to sleep now will see the repercussions later in life.

Get ahead of the pack and join the movement of being *energized,* while the world watches you take the lead.

MICHAEL SAYS...

Wait for it.

During my cardiac event, coming back to a conscious state without vision was terrifying. I know there was a lot of noise around me that day, but all I could hear was my own voice, then my friend's. I barely whispered, with a tear rolling down my cheek, "I can't see."

He immediately and confidently said, "Wait for it."

That was the longest five seconds of my life.

I knew something had gone terribly wrong, that my friend was holding my head, I was blind, and I felt sick. I heard my wife's voice sounding scared but did not know what she was saying. Then all of a sudden, it was like someone hit the fast-forward button on my brain and in the utter blackness of my vision, I saw a very small white and green light and then everything exploded.

There were wires and tubes all over me, and then it went dark and happened again, and then again. It's completely weird when your body just takes over your brain and says *stop;* you really have no choice. So I did.

I stopped a lot of things and started seeking balance.

Having more stress does not mean I need more exercise. It means I need the right exercise, with the right sleep, and the right nutrition. It's all about balance for reduced stress, endless energy, and happiness.

I took my friend's advice, and I have learned to wait for it.

Something else has started happening with me these days. My Wolfishness seems to be moving earlier! This is one of the many chronotypical transitions we humans go through in this thing called life. My chronolongevity is shifting, but I am prepared to seek my new personal balance. Oh, the joys of getting older...

As we write this book, the world is in a very unbalanced place. While some of us will do things to change the world, we can all do something to change our own world, and create our own balance. Energizing is a quest that makes us stronger as we go...

Acknowledgments

From Michael

Stacey Griffith: Well, this certainly has been like Mr. Toad's Wild Ride at Disney. I love you, your energy, and all your insanity. You are the embodiment of *energized* — thank you for teaching me how to get there.

Valerie Frankel: Another one down; as I tell you frequently, you are the only person I will ever trust to tell my thoughts and stories. I can't describe our relationship to anyone, but you get me and I am so grateful that you do and to have you in my life. You are just a great human; thank you for all your help.

Tracy Behar: I can't thank you enough for having faith in our second project together. It has certainly been a different flavor this time around, and I can't thank you enough for your belief in my work and for realizing its importance to the world.

Ian Straus: Thank you, Ian, for *all* the help — in this book and hopefully many more to come.

Alex Glass: You are a super-agent. It's really true. I sing your praises all the time to all my friends and colleagues. And I'm grateful for your tireless ability to deal with my drama and still laugh with me about our kids. You are part of my inner circle. Love you, man.

Maggie Rosenberg: Thank you so much for your awesome illustrations.

Everyone at Little, Brown: Thank you for honoring me with faith in my work. I do not take that commitment lightly. I will show my gratitude by helping get this book in the hands of as many people as I possibly can.

Dave Lakhani: Thank you for creating a safe place so I could have the confidence to take the biggest risks of my life, and for catching me when I fail. Not only have you helped me build a business that helps millions of people every day, but you are teaching me how to be a better human.

Becky Johnston: What can I say? Every person who meets you tells me the same thing: "I wish I had Becky in my life." I tell them all the same thing: "She is one in a million, and back off, she is with the Sleep Doctor." I am grateful for your constant positivity and for helping me not freak out when I get lost, frustrated, and upset. I could not help so many people if I did not have you helping me. Thank you for being willing to deal with my intensity and for helping me manage it better.

The team at Bold Approach: Jacklyn—I know we just started working together, but I am stoked so far and look forward to great projects together. And to all the staff and writers who help bring the Sleep Doctor to life every day, I am so grateful to have you on my team.

Graham Purdy, at Hertz, Lichtenstein, Young, and Polk: Normally, I'm not too fond of lawyers (other than the one I am married to), but, Graham, *you* are a wonderful exception to that rule. Your humor, kindness, and attention to detail are unparalleled, as is true of so many people in your firm. I hope one day that we get to hang out as friends and enjoy life together.

Steven Lockley, PhD: Steven, thank you for continuing to open my eyes to the world of circadian rhythms. Your research is foundational to the science, and without it, this type of book could have never been written. Thank you for all your teachings, your challenges, and the pints of beer we like to share on occasion.

Mickey Beyer-Clausen: I swear we are brothers. In the last few years your friendship has been one of the most important aspects supporting my confidence, balance, and growth. I am not sure I can pay you a higher compliment.

Joe Polish: You may be on sabbatical, but you are still in my head and heart. I hope you find what you are looking for, and if I can help, I will.

Arianna Huffington: Thank you for keeping the conversation going about sleep. I value our relationship, and while we do not get to talk as much anymore, your spirit inspires me in many ways. Thank you.

All the amazing scientists who have circadian rhythms as their research interest: Without you, this book would never have been written. I am excited to get it to the masses and to help them. All your work is listed in our notes section, and we want to thank each and every scientist for their contribution and acknowledge all their efforts to help make this book possible.

Linus, Jacob, Carl, Sondra, Kris, Nick, Tim, Atman, Astrid, and all the amazing friends I have at Hästens: Thank you for seeing the value in sleep science in the world of ultimate luxury. You guys are *fire*!

Jeremy Carr and Yin Yu at Rested Health: You two are showing me the path into the future. While I admit there has been quite a bit of kicking and screaming going on (from me), I want to seriously thank you both. I feel that our relationship has forced me to see where sleep is really going, and how I can be a part of that journey. I seriously cannot thank you both enough for continuing to educate, work, and hang with me. Growth hurts; it's simply a fact. Thanks for making it hurt a little less.

Between books I joined a men's group that has changed my life and brought me into a better place. While I feel that the *entire* group of METAL men are truly an important part of my life now and will continue to be so, I wanted to highlight a few of the "Merry Men" I like to hang out with:

My Badass Breathing Brothers: Neil Cannon (our usually balanced and awesome breathing leader), Chuck, David S, Keith M, Ed McG, Harish R, CK, Ian, Nick, Stanley, Ryan, Jon S, Rick B, Dr. Fred, Sean, Clay, Michael L, Sam M, Jason H, Elliot, and anyone I forgot; and to anyone who is new to joining, see you at 7:35 a.m. PST every day, bro!

Special thanks to Kurtis Lee: You started me down the breathwork path, and I cannot thank you enough. I had no idea how powerful this could be and how important. Also special thanks to my two protectors during my first experience: Emilio and

Sanyika, who created my safe space to explore myself and seek peace. Thank you, brothers.

To the Crypto Round Table: Teddy, Ron, Ethan Z, Josh, Austin, and all my Crypto bros! We are going to *crush it,* or at least have a lot of fun trying.

To the High Finance Group: Quigley, Ken R, this is priceless.

To the Foundry: Ian, Will, and Sam—you fuckers are next level.

To my motivators: Sanyika, Emilio (and Rachel), Mighty Paul, Lanre, Issac, Edwin, and Richard Burke, each of you helps me in a different way, none of you know it, but my sincere thanks goes to you all.

To my dear friend and hypnotherapist, Ken Dubner: Thank you for getting me to the place where I could be open to learning the meaning of the word "brotherhood." And remember, "They are coming from the sewer! Watch out!"

Kenny Rutkowski: I love you. When Jimmy K introduced us, I had no idea you would play such a big role in my life. Thank you for teaching me the meaning of brotherhood. I will forever be in your debt and will represent your vision of heart-centered kind men who kick ass for as long as I live, brother.

To my business partners who have helped in so many ways:

Praesidium: Seth, Skip, Tyler, Jim, Dustin, Doug, Paul, Palo, Giangi, Adam, Stephanie, and all the engineers and staff who have made our future together brighter and brighter.

Purple: Misty, Joe, Russ, and all the people I am meeting now and will work with in the future at Purple, we are going to freaking *crush it.*

The Kryo Team: Todd, I continue to be excited to teach the world about how thermoregulation can be a key driver for sleep.
Thank you for allowing me to represent your company.

Matt and Stacey at Arcadia: Looking forward to surprising the world together!

To my Private Tribe: Thank you for being there for me in so many ways:

Dr. Halland Chen and Talia: Let's be honest, you have seen some things that no one else has; thank you for keeping it that way! *Ha!* Love you guys.

Serenna Poon: You seriously are *the rock star.* I honestly enjoy everything you do, and you do it with such an amazing amount of class and sophistication, and to top it all off, you are a great human. You are a super-special person to me, you know that, and I am honored to be your friend.

Dr. Nicole Burkens: Wow, life is fucking crazy! Here is to your continued success and health.

Jimmy Kwik: Always great to be in your company, my friend; happy to be your wingman, bro!

Mike and Lydia Freed: Thank you so much for sharing Post Ranch with me. I realize what a special place it is, and I feel grateful to be a part of it.

Kari at Nutritious Life: What can I say, *clubhouse*? Hilarious. Love that we have reconnected and are having fun.

Jason and Colleen at MBG: My two new compadres in health. I really like what you guys are about; I'm stoked to help a lot of people.

Vishen at Mindvalley: V, I must say I owe a big thank-you to you, my friend. You probably were the very beginning of my journey into many things more holistic, alternative, and fun.

Shaahin Cheyene: Thank you for everything—sponsorship, friendship, and all the ships we will sail together!

Nathaniel G: We are a bit different than the average bear, and I like it that way!

To my Amazing Celebrity Sleepers:

Steve Aoki: Bro, Pokémon, Thanksgiving, and your mom's cookin'. This year Carson wants to jump off the roof!

P-dog and Carter Reum: Looking forward to one day shaking it up! Thanks for all your support.

Carson Daly: CD, what can I say? You have always been an upstanding guy, and I am so honored to have been there to help.

My Amazing Sleep Colleagues:

Dr. Michael Grandner: Looking forward to reading *your* book, my friend.

Dr. Wendy Troxel: Thank you for all your help with TedX and for your friendship. Always happy to support your efforts.

Dr. Heidi Hanna: You and Mark are *awesome,* and I am excited to continue to work together on all our fun stuff.

From Stacey

Dr. Michael Breus, I can't thank you enough times for receiving my energy in the way I believed we would collaborate. You are one of the busiest humans I've ever met, and the fact that you found a way to write a book with *me,* during this time of life, makes me smile from ear to ear. Our two Manhattans were destiny from the beginning.

Val Frankel, our time together through this book was so special. Thank you for delivering this project with us in a way that was authentic, considerate, and timely. You are a very talented human. It is something so hard to find, and yet so much appreciated.

Alex Glass, "agent" is not enough of a word for you. Michael was right—you are the best. Thank you for rolling calls in between being a great dad and husband.

Tracy Behar and Ian Straus, thank you for believing in us, especially when we, or should I say *I,* kept you awake late nights with plugs, lightning bolts, and cover changes.

Kathy Gordon, I'm not sure this entire project could have happened without you. Somehow, some way, you always, without fail, find a way to make me shine. Your knowledge and artistry in publishing completely blow my mind. Legacy Lit just won the lotto in you...

Mom, you always know how to talk me through all my worries, fears, and self-doubts. Thank you for being my biggest fan in this big world full of bullies!

Tiff, Griff, and Emmalyn, I can't count the number of hugs we missed out on during the last two years this book was in motion, but if Dad were here to see us through it, you know he would still expect more. Jules, thanks for holding down the 555 for us all over the last four decades, STAV rules!

Sandy, Tiff, and Koko, I will never forget how much you were there for Gramma Stella. From when she was ninety to one hundred years old, you guys made her life amazing, and the hundredth party for her at Soul Seattle will be one we'll never forget. She and Gary are throwing us another book party up there with Lo, Ed, and Seann.

Michelle Smith, I have my entire life ahead because of you; finding you halfway through it makes getting to the end of it completely whole!

Liz Moran, PT maven and kinesiology queen, I don't know if it's my mind or body that owes you the thank-you, but without you, both things would be broken toys.

ALF, for giving me a sanctuary out east to write with total clarity, in the midst of a global pandemic.

Rain and Barb, my dynamic duo that has nothin' but net always...

Sarah Wragge, my podcast and nutritionist superhero who does it all, from Jersey. LOL.

SoulCycle, my platform of positivity, strength, community, and love. It has been a journey over the first fifteen years that has changed my life and continues to excite me at every Two Turns from Zero.

All of my Soul Squad riders, some of whom have become friends over the last fifteen years, and some of whom have nicknames that would require their own book, you are the fire that wakes me up every day! Classes with you are, to me, the "meeting" I needed to live a clean-livin' life. I may forget names, but I will always remember your faces!

Neil, David, "H & G," for always being the role models of modern-day parenting.

Without constant deliveries from Akeso Water and Zing Bar I would be dehydrated and hungry!

Thank you, @LYMBR, for the stretches after long writing edits and long days teaching!

To all the people who brought light to the world during the pandemic — our health care workers, schoolteachers, fitness professionals, essential workers in all industries, parents, everyone who made our new normal *normal* — as a teacher/instructor of fitness for more than thirty years, I learned from all of you that giving to the human spirit is the ultimate contribution.

Thank you, @Instagram, for providing the global platform to do so.

Notes

Introduction: Our Lightning Strike

1. Potter GD, Skene DJ, Arendt J, Cade JE, Grant PJ, Hardie LJ. Circadian rhythm and sleep disruption: Causes, metabolic consequences, and countermeasures. *Endocr Rev.* 2016;37(6):584–608. doi:10.1210/er.2016-1083.
2. Walker WH, Walton JC, DeVries AC, et al. Circadian rhythm disruption and mental health. *Transl Psychiatry.* 2020;10(28).
3. Kitchen GB, et al. The clock gene *Bmal1* inhibits macrophage motility, phagocytosis, and impairs defense against pneumonia. *PNAS.* 2020 Jan;117(3):1543–1551.

Chapter 1: Exhausted to Energized: The Energy Scale

1. Borg GA. Psychophysical bases of perceived exertion. *Med Sci Sports Exerc.* 1982;14(5): 377–381.
2. Sleep in America Poll 2020. National Sleep Foundation / Langer Research Associates, 2020. sleepfoundation.org/sites/default/files/2020-03/SIA%202020%20Q1%20 Report.pdf.
3. Mehta RK, et al. Relationship between BMI and fatigability is task dependent. *Hum Factors.* 2017 Aug.
4. Nedeltcheva AV, Kilkus JM, Imperial J, Schoeller DA, Penev PD. Insufficient sleep undermines dietary efforts to reduce adiposity. *Ann Intern Med.* 2010;153(7):435–441. doi:10.7326/0003-4819-153-7-201010050-00006.
5. Armstrong LE, et al. Mild dehydration affects mood in healthy young women. *J Nutr.* 2011 Dec.

Chapter 2: What's Your Power Profile? Body Type and Chronotype

1. Sheldon WH. *Atlas of Men: A Guide for Somatotyping the Adult Male at All Ages.* New York: Harper; 1954.
2. Randler C, Schredl M, Göritz AS. Chronotype, sleep behavior, and the big five personality factors. *SAGE Open.* 2017 Jul. doi:10.1177/2158244017728321.
3. Suh S, Yang HC, Kim N, et al. Chronotype differences in health behaviors and

health-related quality of life: A population-based study among aged and older adults. *Behav Sleep Med.* 2017 Sep–Oct;15(5):361–376.

4. Preckel F, Lipnevich AA, Schneider S, Roberts RD. Chronotype, cognitive abilities, and academic achievement: A meta-analytic investigation. *Learn Individ Differ.* 2011; 21(5):483–492.

5. Jankowski K. Morningness/eveningness and satisfaction with life in a Polish sample. *Chronobiol Int.* 2012;29:780–785.

6. Koleva M, Nacheva A, Boev M. Somatotype and disease prevalence in adults. *Rev Environ Health.* 2002 Jan–Mar;17(1):65–84.

7. Koleva M, Nacheva A, Boev M. Somatotype, nutrition, and obesity. *Rev Environ Health.* 2000 Oct–Dec;15(4):389–398.

8. Lizana PA, Olivares R, Berral FJ. Somatotype tendency in Chilean adolescents from Valparaíso: Review from 1979 to 2011. *Nutr Hosp.* 2014 Dec 17;31(3):1034–1043.

9. Rajkumar RV. Endomorphy dominance among non-athlete population in all the ranges of body mass index. *Int J Physiother Res.* 2015;3:1068–1074. doi:10.16965/ijpr.2015.139.

Chapter 3: Resting Energy

1. Smith TJ, et al. Impact of sleep restriction on local immune response and skin barrier restoration with and without "multinutrient" nutrition and intervention. *J Appl Physiol.* 2018.

2. Fultz NE, et al. Coupled electrophysiological, hemodynamic, and cerebrospinal fluid oscillations in human sleep. *Science.* 2019.

3. Jessen NA, Munk AS, Lundgaard I, Nedergaard M. The glymphatic system: A beginner's guide. *Neurochem Res.* 2015;40(12):2583–2599. doi:10.1007/s11064-015-1581-6.

4. Paller KA, Creery JD, Schechtman E. Memory and sleep: How sleep cognition can change the waking mind for the better. *Annu Rev Psychol.* 2021;72:123–150.

5. Gharib SA, et al. Transcriptional signatures of sleep duration discordance in monozygotic twins. *Sleep.* 2017 Jan.

6. Mazzoccoli G, et al. Comparison of whole body circadian phase evaluated from melatonin and cortisol secretion profiles in healthy humans. *Biomed Aging Pathol.* 2011 Apr–Jun;1(2): 112–122.

7. Smith MG, Wusk GC, Nasrini J, et al. Effects of six weeks of chronic sleep restriction with weekend recovery on cognitive performance and wellbeing in high-performing adults. *Sleep.* 2021 Feb. doi:10.1093/sleep/zsab051.

8. Nutt D, et al. Sleep disorders as core symptom of depression. *Dialogues Clin Neurosci.* 2008.

9. Better Sleep Council, "Sleeping Together Can Be a Nightmare for Couples," September 2012.

10. Blume C, Garbazza C, Spitschan M. Effects of light on human circadian rhythms, sleep and mood. *Somnologie (Berl).* 2019;23(3):147–156. doi:10.1007/s11818-019-00215-x.

11. Rosinger AY, Chang A-M, Buxton OM, Li J, Wu S, Gao X. Short sleep duration is associated with inadequate hydration: Cross-cultural evidence from US and Chinese adults. *Sleep.* 2019 Feb;42(2).

12. Fonsêca NT, Santos IR, Fernandes V, Fernandes VAT, Lopes VCD, Luis VFO. Excessive

daytime sleepiness in patients with chronic kidney disease undergone hemodialysis. *Fisioter Mov.* 2014;27(4):653–660.

13. Mooventhan A, Nivethitha L. Scientific evidence-based effects of hydrotherapy on various systems of the body. *N Am J Med Sci.* 2014;6(5):199–209. doi:10.4103/1947-2714.132935.

14. Taheri M, Irandoust K. Morning exercise improves cognitive performance decrements induced by partial sleep deprivation in elite athletes. *Biol Rhythm Res.* 2020;51.

15. Wheeler MJ, Green DJ, Ellis KA, et al. Distinct effects of acute exercise and breaks in sitting on working memory and executive function in older adults: A three-arm, randomised cross-over trial to evaluate the effects of exercise with and without breaks in sitting on cognition. *Br J Sports Med.* 2020;54:776–781.

16. Morita Y, Sasai-Sakuma T, Inoue Y. Effects of acute morning and evening exercise on subjective and objective sleep quality in older individuals with insomnia. *Sleep Med.* 2017 Jun;34:200–208.

17. National Coffee Association. NCA releases Atlas of American Coffee. 2020 Mar 26. ncausa.org/Newsroom/NCA-releases-Atlas-of-American-Coffee.

18. Ágoston C, Urbán R, Rigó A, Griffiths MD, Demetrovics Z. Morningness-eveningness and caffeine consumption: A largescale path-analysis study. *Chronobiol Int.* 2019 Sep;36(9):1301–1309. doi:10.1080/07420528.2019.1624372.

19. Burke TM, et al. Effects of caffeine on the human circadian clock in vivo and in vitro. *Sci Transl Med.* 2015 Sep 16;7(305):305ra146.

20. Chaudhary NS, Grandner MA, Jackson NJ, Chakravorty S. Caffeine consumption, insomnia, and sleep duration: Results from a nationally representative sample. *Nutrition.* 2016;32(11–12):1193–1199. doi:10.1016/j.nut.2016.04.005.

21. Drake C, Roehrs T, Shambroom J, Roth T. Caffeine effects on sleep taken 0, 3, or 6 hours before going to bed. *J Clin Sleep Med.* 2013 Nov 15;9(11):1195–1200. doi:10.5664/jcsm.3170.

22. Gabel V, Reichert CF, Maire M, et al. Differential impact in young and older individuals of blue-enriched white light on circadian physiology and alertness during sustained wakefulness. *Sci Rep.* 2017;7(1):7620.

23. University of Haifa. Blue light emitted by screens damages our sleep, study suggests. *ScienceDaily.* 2017 Aug 22.

24. Barcelona Institute for Global Health (ISGlobal). Study links night exposure to blue light with breast and prostate cancer: Researchers used images taken by astronauts to evaluate outdoor lighting in Madrid and Barcelona. *ScienceDaily.* 2018 Apr 25.

25. Gabel V, Reichert CF, Maire M, et al. Differential impact in young and older individuals of blue-enriched white light on circadian physiology and alertness during sustained wakefulness. *Sci Rep.* 2017;7(1):7620.

26. Masís-Vargas A, Hicks D, Kalsbeek A, Mendoza J. Blue light at night acutely impairs glucose tolerance and increases sugar intake in the diurnal rodent Arvicanthis ansorgei in a sex-dependent manner. *Physiol Rep.* 2019;7(20):e14257. doi:10.14814/phy2.14257.

27. Esaki Y, Kitajima T, Ito Y, et al. Wearing blue light-blocking glasses in the evening advances circadian rhythms in the patients with delayed sleep phase disorder: An open-label trial. *Chronobiol Int.* 2016;33(8):1037–1044.

28. Mortazavi SAR, Parhoodeh S, Hosseini MA, et al. Blocking short-wavelength component of the visible light emitted by smartphones' screens improves human sleep quality. *J Biomed Phys Eng.* 2018;8(4):375–380.

29. Nagare R, Plitnick B, Figueiro M. Does the iPad Night Shift mode reduce melatonin suppression? *Light Res Technol.* 2019;51(3):373–383.

30. Chan, JM, et al. The acute effects of alcohol on sleep electroencephalogram power spectrum in late adolescence. *Alcohol Clin Exp Res.* 2015 Feb;39(2).

31. Ruby CL, Brager AJ, DePaul MA, Prosser RA, Glass JD. Chronic ethanol attenuates circadian photic phase resetting and alters nocturnal activity patterns in the hamster. *Am J Physiol Regul Integr Comp Physiol.* 2009 Sep;297(3):R729–R737.

32. Rupp TL, Acebo C, Carskadon MA. Evening alcohol suppresses salivary melatonin in young adults. *Chronobiol Int.* 2007;24(3):463–470.

33. Thakkar MM, Sharma R, Sahota P. Alcohol disrupts sleep homeostasis. *Alcohol.* 2015;49(4):299–310.

34. Pietilä J, Helander E, Korhonen I, Myllymäki T, Kujala UM, Lindholm H. Acute effect of alcohol intake on cardiovascular autonomic regulation during the first hours of sleep in a large real-world sample of Finnish employees: Observational study. *JMIR Ment Health.* 2018;5(1):e23.

35. Roehrs T, Roth T. Sleep, sleepiness, sleep disorders and alcohol use and abuse. *Sleep Med Rev.* 2001 Aug;5(4):287–297.

36. Ebrahim IO, Shapiro CM, Williams AJ, Fenwick PB. Alcohol and sleep I: Effects on normal sleep. *Alcohol Clin Exp Res.* 2013 Apr;37(4):539–549. doi:10.1111/acer.12006.

37. Mograss M, Crosetta M, Abi-Jaoude J, et al. Exercising before a nap benefits memory better than napping or exercising alone. *Sleep.* 2020 Sep 14;43(9).

38. Moser RS, et al. Efficacy of immediate and delayed cognitive and physical rest for treatment of sports-related concussion. *J Pediatr.* 2012 Nov 1;161(5):922–926.

39. Tarleton EK, Littenberg B, MacLean CD, Kennedy AG, Daley C. Role of magnesium supplementation in the treatment of depression: A randomized clinical trial. *PLOS One.* 2017;12(6):e0180067.

40. Möykkynen T, Uusi-Oukari M, Heikkilä J, Lovinger DM, Lüddens H, Korpi ER. Magnesium potentiation of the function of native and recombinant GABA(A) receptors. *Neuroreport.* 2001 Jul 20;12(10):2175–2179. doi:10.1097/00001756-200107200-00026.

41. Piovezan RD, Hirotsu C, Feres MC, et al. Obstructive sleep apnea and objective short sleep duration are independently associated with the risk of serum vitamin D deficiency. *PLOS One.* 2017 Jul 7;12(7):e0180901.

42. Gao Q, Kou T, Zhuang B, Ren Y, Dong X, Wang Q. The association between vitamin D deficiency and sleep disorders: A systematic review and meta-analysis. *Nutrients.* 2018;10(10):1395.

43. Aspy DJ, et al. Effects of vitamin B6 (pyridoxine) and a B complex preparation on dreaming and sleep. *Percept Mot Skills.* 2018 Apr 17;125(3).

44. Kripke DF. Hypnotic drug risks of mortality, infection, depression, and cancer: But lack of benefit. *F1000Res.* 2016;5:918. doi:10.12688/f1000research.8729.3.

45. Weaver MF. Prescription sedative misuse and abuse. *Yale J Biol Med.* 2015;88(3): 247–256.

46. Kripke DF. Hypnotic drug risks of mortality, infection, depression, and cancer: But lack of benefit. *F1000Res.* 2016;5:918. doi:10.12688/f1000research.8729.3.

47. Kripke DF, Langer RD, Kline LE. Hypnotics' association with mortality or cancer: A matched cohort study. *BMJ Open.* 2012;2:e000850.

48. Sateia MJ, Buysse DJ, Krystal AD, Neubauer DN, Heald JL. Clinical practice guideline for the pharmacologic treatment of chronic insomnia in adults: An American Academy of Sleep Medicine clinical practice guideline. *J Clin Sleep Med.* 2017 Feb 15;13(2): 307–349.

49. Leng Y, et al. Sleep medication use and risk of dementia in a biracial cohort of older adults. Presented at: Alzheimer's Association International Conference; July 14–18, 2019; Los Angeles.

50. Jacobs GD, Pace-Schott EF, Stickgold R, Otto MW. Cognitive behavior therapy and pharmacotherapy for insomnia: A randomized controlled trial and direct comparison. *Arch Intern Med.* 2004 Sep 27;164(17):1888–1896. doi:10.1001/archinte.164.17.1888.

51. Tringale R, Jensen C. Cannabis and insomnia. *O'Shaughnessy's.* Autumn 2011.

52. Schierenbeck T, Riemann D, Berger M, Hornyak M. Effect of illicit recreational drugs upon sleep: Cocaine, ecstasy and marijuana. *Sleep Med Rev.* 2008 Oct;12(5):381–389.

53. Shannon S, Lewis N, Lee H, Hughes S. Cannabidiol in anxiety and sleep: A large case series. *Perm J.* 2019;23:18-041. doi:10.7812/TPP/18-041.

54. Peana AT, Rubattu P, Piga GG, et al. Involvement of adenosine A1 and A2A receptors in (-)-linalool-induced antinociception. *Life Sci.* 2006 Apr 18;78(21):2471–2474.

Chapter 4: Moving Energy

1. Yang L, Cao C, Kantor ED, et al. Trends in sedentary behavior among the US population, 2001–2016. *JAMA.* 2019;321(16):1587–1597. doi:10.1001/jama.2019.3636.

2. Janssen I, et al. A systematic review of compositional data analysis studies examining associations between sleep, sedentary behaviour, and physical activity with health outcomes in adults. *Appl Physiol Nutr Metab.* 2020;45(10 (Suppl. 2)):S248–S257.

3. World Health Organization. Physical activity. 2020 Nov 6. who.int/news-room/fact-sheets/detail/physical-activity.

4. Willis EA, et al. The effects of exercise session timing on weight loss and components of energy balance. *Int J Obes (Lond).* 2020 Jan;44(1):114–124. doi:10.1038/s41366-019-0409-x.

5. Asher G, et al. Physiology and molecular dissection of daily variance in exercise capacity. *Cell Metab.* 2019 Apr.

6. Stutz J, Eiholzer R, Spengler CM. Effects of evening exercise on sleep in healthy participants: A systematic review and meta-analysis. *Sports Med.* 2019 Feb;49(2):269–287.

7. Hunt MG, Marx R, Lipson C, Young J. No more FOMO: Limiting social media decreases loneliness and depression. *J Soc Clin Psychol.* 2018;37(10):751–768.

8. Ha AS, Ng JYY. Rope skipping increases bone mineral density at calcanei of pubertal girls in Hong Kong: A quasi-experimental investigation. *PLOS One.* 2017;12(12):e0189085.

9. Baker JA. Comparison of rope skipping and jogging as methods of improving cardiovascular efficiency of college men. *Res Q.* 1968;39(2):240–243.

10. Holloszy JO. The biology of aging. *Mayo Clin Proc.* 2000;75(Suppl):S3–S8.

11. Volpi E, Nazemi R, Fujita S. Muscle tissue changes with aging. *Curr Opin Clin Nutr Metab Care.* 2004;7(4):405–410.

12. Beavers KM, Ambrosius WT, Rejeski WJ, et al. Effect of exercise type during intentional weight loss on body composition in older adults with obesity. *Obesity (Silver Spring).* 2017 Nov;25(11):1823–1829.

Chapter 5: Eating Energy

1. St-Onge MP, Roberts A, Shechter A, Choudhury AR. Fiber and saturated fat are associated with sleep arousals and slow wave sleep. *J Clin Sleep Med.* 2016 Jan;12(1):19–24.

2. Wijngaarden MA, van der Zon GC, van Dijk KW, Pijl H, Guigas B. Effects of prolonged fasting on AMPK signaling, gene expression, and mitochondrial respiratory chain content in skeletal muscle from lean and obese individuals. *Am J Physiol Endocrinol Metab.* 2013;304(9):E1012–E1021.

3. De Cabo R, Mattson MP. Effects of intermittent fasting on health, aging and disease. *N Engl J Med.* 2019 Dec 26;381(26):2541–2551.

4. Alirezaei M, Kemball CC, Flynn CT, Wood MR, Whitton JL, Kiosses WB. Short-term fasting induces profound neuronal autophagy. *Autophagy.* 2010;6(6):702–710.

5. Liu B, et al. Intermittent fasting increases energy expenditure and promotes adipose tissue browning in mice. *Nutrition.* 2019 Oct;66:38–43.

6. Hatori M, et al. Time-restricted feeding without reducing calorie intake prevents metabolic disease in mice fed a high-fat diet. *Cell Metab.* 2012 Jun;15(6):848–860.

7. Public Health England. National Diet and Nutrition Survey. 2016 Sep 9. gov.uk/government/collections/national-diet-and-nutrition-survey.

8. Bergendahl M, Evans WS, Pastor C, Patel A, Iranmanesh A, Veldhuis JD. Short-term fasting suppresses leptin and (conversely) activates disorderly growth hormone secretion in midluteal phase women—a clinical research center study. *J Clin Endocrinol Metab.* 1999 Mar 1;84(3):883–894.

9. Espelund U, Hansen TK, Højlund K, et al. Fasting unmasks a strong inverse association between ghrelin and cortisol in serum: Studies in obese and normal-weight subjects. *J Clin Endocrinol Metab.* 2005 Feb;90(2):741–746. doi:10.1210/jc.2004-0604.

10. Mercola J. *Fat for Fuel.* Carlsbad, CA: Hay House; 2017.

11. Betts JA, Richardson JD, Chowdhury EA, Holman GD, Tsintzas K, Thompson D. The causal role of breakfast in energy balance and health: A randomized controlled trial in lean adults. *Am J Clin Nutr.* 2014 Aug;100(2):539–547.

12. Jakubowicz D, Barnea M, Wainstein J, Froy O. High caloric intake at breakfast vs. dinner differentially influences weight loss of overweight and obese women. *Obesity (Silver Spring).* 2013 Dec;21(12):2504–2512. doi:10.1002/oby.20460.

13. Paoli A, Tinsley G, Bianco A, Moro T. The influence of meal frequency and timing on health in humans: The role of fasting. *Nutrients.* 2019;11(4):719.

14. Sutton EF, et al. Early time-restricted feeding improves insulin sensitivity, blood pressure, and oxidative stress even without weight loss in men with prediabetes. *Cell Metab.* 2018 Jun 5;27(6):1212–1221.e3.

15. Qasrawi SO, Pandi-Perumal SR, BaHammam AS. The effect of intermittent fasting during Ramadan on sleep, sleepiness, cognitive function, and circadian rhythm. *Sleep Breath.* 2017 Sep;21(3):577–586.

16. Crispin CA, et al. Relationship between food intake and sleep pattern in healthy individuals. *J Clin Sleep Med.* 2011 Dec 15;7(6).

17. Gu C, Brereton N, Schweitzer A, et al. Metabolic effects of late dinner in healthy volunteers—a randomized crossover clinical trial. *J Clin Endocrinol Metab.* 2020 Aug;105(8):2789–2802.

Chapter 6: Emotional Energy

1. Mauss IB, Tamir M, Anderson CL, Savino NS. Can seeking happiness make people unhappy? [corrected] Paradoxical effects of valuing happiness [published correction appears in *Emotion*. 2011 Aug;11(4):767]. *Emotion*. 2011;11(4):807–815.

2. Stolarski M, Jankowski KS, Matthews G, Kawalerczyk J. Wise "birds" follow their clock: The role of emotional intelligence and morningness-eveningness in diurnal regulation of mood. *Chronobiol Int*. 2016;33(1):51–63.

3. Yim J. Therapeutic benefits of laughter in mental health: A theoretical review. *Tohoku J Exp Med*. 2016 Jul;239(3):243–249.

4. Idris DNT, Astarani K, Mahanani S. The comparison between the effectiveness of laughter therapy and progressive muscle relaxation therapy towards insomnia in elderly community at St. Yoseph Kediri Nursing Home. *Indian J Public Health Res Dev*. 2020;11(9):218–225.

5. Kimata H. Laughter elevates the levels of breast-milk melatonin. *J Psychosom Res*. 2007 Jun;62(6):699–702.

6. Salimpoor V, Benovoy M, Larcher K, et al. Anatomically distinct dopamine release during anticipation and experience of peak emotion to music. *Nat Neurosci*. 2011; 14:257–262.

7. Thoma MV, La Marca R, Brönnimann R, Finkel L, Ehlert U, Nater UM. The effect of music on the human stress response. *PLOS One*. 2013;8(8):e70156.

8. Leubner D, Hinterberger T. Reviewing the effectiveness of music interventions in treating depression. *Front Psychol*. 2017;8:1109.

9. Bottiroli S, Rosi A, Russo R, Vecchi T, Cavallini E. The cognitive effects of listening to background music on older adults: Processing speed improves with upbeat music, while memory seems to benefit from both upbeat and downbeat music. *Front Aging Neurosci*. 2014;6:284.

10. Alcântara-Silva TR, de Freitas-Junior R, Freitas NMA, et al. Music therapy reduces radiotherapy-induced fatigue in patients with breast or gynecological cancer: A randomized trial. *Integr Cancer Ther*. 2018 Sep:628–635.

11. Brown L, Houston EE, Amonoo HL, et al. Is self-compassion associated with sleep quality? A meta-analysis. *Mindfulness*. 2020.

12. Klimecki OM, Leiberg S, Lamm C, Singer T. Functional neural plasticity and associated changes in positive affect after compassion training. *Cereb Cortex*. 2013 July; 23(7):1552–1561.

13. Fuegen K, Breitenbecher KH. Walking and being outdoors in nature increase positive affect and energy. *Ecopsychology*. 2018 Mar;10(1):14–25.

14. Slawinska M, et al. Effects of chronotype and time of day on mood responses to Cross-Fit training. *Chronobiol Int*. 2019 Feb;36(2):237–249.

15. Laborde S, et al. Chronotype, sports participation, and positive personality-train-like individual differences. *Chronobiol Int*. 2015;32(7):942–951.

16. Kalmbach DA, Arnedt JT, Pillai V, Ciesla JA. Sex and sleep. *J Sex Med*. 2015; 12:1221–1232.

17. Jankowski KS, Díaz-Morales JF, Randler C. Chronotype, gender, and time for sex. *Chronobiol Int*. 2014 Oct;31(8):911–916. doi:10.3109/07420528.2014.925470.

18. Chellappa SL, Morris CJ, Scheer FAJL. Circadian misalignment increases mood vulnerability in simulated shift work. *Sci Rep.* 2020;10:18614.

19. Taylor BJ, Hasler BP. Chronotype and mental health: Recent advances. *Curr Psychiatry Rep.* 2018.

20. Glosemeyer RW, Diekelmann S, Cassel W, et al. Selective suppression of rapid eye movement sleep increases next-day negative affect and amygdala responses to social exclusion. *Sci Rep.* 2020;10:17325.

21. Van der Helm E, Gujar N, Walker MP. Sleep deprivation impairs the accurate recognition of human emotions. *Sleep.* 2010 Mar;33(3):335–342.

22. Guadagni V, Burles F, Ferrara M, Iaria G. The effects of sleep deprivation on emotional empathy. *J Sleep Res.* 2014;23:657–663.

23. Nota JB, Coles ME. Shorter sleep duration and longer sleep onset latency are related to difficulty disengaging attention from negative emotional images in individuals with elevated transdiagnostic repetitive negative thinking. *J Behav Ther Exp Psychiatry.* 2018 Mar;58:112–122.

24. Goldstein AN, Greer SM, Saletin JM, Harvey AG, Nitschke JB, Walker MP. Tired and apprehensive: Anxiety amplifies the impact of sleep loss on aversive brain anticipation. *J Neurosci.* 2013;33(26):10607–10615.

25. Demirer I, Erol S. The relationship between university students' physical activity levels, insomnia and psychological well-being. *J Psychiatric Nurs.* 2020;11(3):201–211.

Chapter 7: The Medium Bear Power Protocol

1. Zahrt OH, Crum AJ. Perceived physical activity and mortality: Evidence from three nationally representative U.S. samples. *Health Psychol.* 2017 Nov;36(11):1017–1025.

2. Blanchfield AW, Hardy J, De Morree HM, Staiano W, Marcora SM. Talking yourself out of exhaustion: The effects of self-talk on endurance performance. *Med Sci Sports Exerc.* 2014;46(5):998–1007.

3. Bolino MC, Hsiung H-H, Harvey J, LePine JA. "Well, I'm tired of tryin'!" Organizational citizenship behavior and citizenship fatigue. *J Appl Psychol.* 2015;100(1):56–74.

4. Chtourou H, et al. The effect of training at a specific time of day: A review. *J Strength Cond Res.* 2012 Jul.

Chapter 9: The Medium Wolf Power Protocol

1. Gates PJ, Albertella L, Copeland J. The effects of cannabinoid administration on sleep: A systematic review of human studies. *Sleep Med Rev.* 2014 Dec;18(6):477–487.

2. Tringale R. Cannabis and insomnia. *O'Shaughnessy's.* Autumn 2011.

3. Chopda GR, Parge V, Thakur GA, Gatley SJ, Makriyannis A, Paronis CA. Tolerance to the diuretic effects of cannabinoids and cross-tolerance to a κ-opioid agonist in THC-treated mice. *J Pharmacol Exp Ther.* 2016;358(2):334–341.

4. Prestifilippo JP, Fernández-Solari J, de la Cal C, et al. Inhibition of salivary secretion by activation of cannabinoid receptors. *Exp Biol Med (Maywood).* 2006 Sep;231(8):1421–1429.

5. Schierenbeck T, Riemann D, Berger M, Hornyak M. Effect of illicit recreational drugs

upon sleep: Cocaine, ecstasy and marijuana. *Sleep Med Rev.* 2008 Oct;12(5):381–389. doi:10.1016/j.smrv.2007.12.004.

6. Wu J. Cannabis, cannabinoid receptors, and endocannabinoid system: Yesterday, today, and tomorrow. *Acta Pharmacol Sin.* 2019;40:297–299.

7. Winiger EA, Ellingson JM, Morrison CL, et al. Sleep deficits and cannabis use behaviors: An analysis of shared genetics using linkage disequilibrium score regression and polygenic risk prediction. *Sleep.* 2020;zsaa188. doi:10.1093/sleep/zsaa188.

8. Conroy DA, Kurth ME, Strong DR, Brower KJ, Stein MD. Marijuana use patterns and sleep among community-based young adults. *J Addict Dis.* 2016;35(2):135–143.

9. Gao Q, Sheng J, Qin S, Zhang L. Chronotypes and affective disorders: A clock for mood? *Brain Science Advances.* 2019;5(3):145–160. doi:10.26599/BSA.2019.9050018.

10. Rynders CA, Thomas EA, Zaman A, Pan Z, Catenacci VA, Melanson EL. Effectiveness of intermittent fasting and time-restricted feeding compared to continuous energy restriction for weight loss. *Nutrients.* 2019;11(10):2442.

11. Leow S, Jackson B, Alderson JA, Guelfi KJ, Dimmock JA. A role for exercise in attenuating unhealthy food consumption in response to stress. *Nutrients.* 2018;10(2):176. doi:10.3390/nu10020176.

12. Ma X, Yue ZQ, Gong ZQ, et al. The effect of diaphragmatic breathing on attention, negative affect and stress in healthy adults. *Front Psychol.* 2017;8:874.

13. Peper E, Harvey R, Hamiel V. Transforming thoughts with postural awareness to increase therapeutic and teaching efficacy. *NeuroRegulation.* 2019;6:153–160. doi:10.15540/nr.6.3.153.

Chapter 10: The Slow Wolf Power Protocol

1. Vetter C, Chang S-C, Devore EE, Rohrer F, Okereke OI, Schernhammer ES. Prospective study of chronotype and incident depression among middle- and older-aged women in the nurses' Health Study II. *J Psychiatr Res.* 2018.

2. Mazri FN, et al. The association between chronotype and dietary pattern among adults: A scoping review. *Int J Environ Res Public Health.* 2019 Nov.

3. McCall B. Hot tubs improve A1c, BMI, and blood pressure in type 2 diabetes. Medscape. 2020 Sep 24. medscape.com/viewarticle/937991.

4. Hoffman BM, Babyak MA, Craighead WE, et al. Exercise and pharmacotherapy in patients with major depression: One-year follow-up of the SMILE study. *Psychosom Med.* 2011 Feb–Mar;73(2):127–133.

Chapter 11: The Fast and Medium Lion Power Protocol

1. Lipnevich AA, Credè M, Hahn E, Spinath FM, Roberts RD, Preckel F. How distinctive are morningness and eveningness from the Big Five factors of personality? A meta-analytic investigation. *J Pers Soc Psychol.* 2017;112(3):491–509.

2. Cadegiani FA, Kater CE. Body composition, metabolism, sleep, psychological and eating patterns of overtraining syndrome: Results of the EROS study (EROS-PROFILE). *J Sports Sci.* 2018 Aug;36(16):1902–1910.

3. Centers for Disease Control and Prevention. Physical activity recommendations

for different age groups. 2021 Apr 13. cdc.gov/physicalactivity/basics/age-chart
.html.

4. Maukonen M, Kanerva N, Partonen T, et al. Chronotype differences in timing of
    energy and macronutrient intakes: A population-based study in adults. *Obesity.*
    2017;25(3):608.

5. Stolarski M, Jankowski KS. Morningness–eveningness and performance-based emo-
    tional intelligence. *Biol Rhythm Res.* 2015;46(3):417–423.

6. Jankowski KS. Morningness/eveningness and satisfaction with life in a Polish sample.
    Chronobiol Int. 2012 Jul;29(6):780–785. doi:10.3109/07420528.2012.685671.

7. Faraut B, Nakib S, Drogou C, et al. Napping reverses the salivary interleukin-6 and uri-
    nary norepinephrine changes induced by sleep restriction. *J Clin Endocrinol Metab.* 2015
    Mar;100(3):E416–E426.

8. Piro C, Fraioli F, Sciarra F, Conti C. Circadian rhythm of plasma testosterone, cortisol
    and gonadotropins in normal male subjects. *J Steroid Biochem.* 1973 May;4(3):321–329.

9. Jankowski KS, Ciarkowska W. Diurnal variation in energetic arousal, tense arousal,
    and hedonic tone in extreme morning and evening types. *Chronobiol Int.* 2008
    Jul;25(4):577–595.

10. Deguchi Y, Miyazaki K. Anti-hyperglycemic and anti-hyperlipidemic effects of guava
    leaf extract. *Nutr Metab (Lond).* 2010;7(9). doi:10.1186/1743-7075-7-9.

11. Mazri FH, Manaf ZA, Shahar S, Mat Ludin AF. The association between chronotype
    and dietary pattern among adults: A scoping review. *Int J Environ Res Public Health.*
    2019;17(1):68.

12. Díaz-Morales J. Morning and evening-types: Exploring their personality styles. *Pers
    Individ Differ.* 2007;43:769–778. doi:10.1016/j.paid.2007.02.002.

Chapter 12: The Slow Lion Power Protocol

1. Pruszczak D, Stolarski M, Jankowski KS. Chronotype and time metaphors: Morning-
    types conceive time as more friendly and less hostile. *Biol Rhythm Res.* 2018;49(3):431–441.

2. Díaz-Morales JF, Ferrari JR, Cohen JR. Indecision and avoidant procrastination: The
    role of morningness-eveningness and time perspective in chronic delay lifestyles. *J Gen
    Psychol.* 2008 Jul;135(3):228–240.

3. Ekelund U, Tarp J, Fagerland MW, et al. Joint associations of accelero-meter measured
    physical activity and sedentary time with all-cause mortality: A harmonised meta-
    analysis in more than 44 000 middle-aged and older individuals. *Br J Sports Med.*
    2020;54:1499–1506.

4. Hurst Y, Fukuda H. Effects of changes in eating speed on obesity in patients with diabe-
    tes: A secondary analysis of longitudinal health check-up data. *BMJ Open.* 2018;8:e019589.

5. Maukonen M, Kanerva N, Partonen T, et al. Chronotype differences in timing of
    energy and macronutrient intakes: A population-based study in adults. *Obesity.* 2017;25:
    608–615.

6. Deguchi Y, Miyazaki K. Anti-hyperglycemic and anti-hyperlipidemic effects of guava
    leaf extract. *Nutr Metab (Lond).* 2010;7(9). doi:10.1186/1743-7075-7-9.

7. Whillans AV, Dunn EW, Sandstrom GM, Dickerson SS, Madden KM. Is spending
    money on others good for your heart? *Health Psychol.* 2016 Jun;35(6):574–583.

Chapter 13: The Fast and Medium Dolphin Power Protocol

1. Kripke DF. Hypnotic drug risks of mortality, infection, depression, and cancer: But lack of benefit. *F1000Res.* 2016;5:918. doi:10.12688/f1000research.8729.3.

2. Jacobs GD, Pace-Schott EF, Stickgold R, Otto MW. Cognitive behavior therapy and pharmacotherapy for insomnia: A randomized controlled trial and direct comparison. *Arch Intern Med.* 2004 Sep 27;164(17):1888–1896. doi:10.1001/archinte.164.17.1888.

3. Day MV, Bobocel DR. The weight of a guilty conscience: Subjective body weight as an embodiment of guilt. *PLOS One.* 2013;8(7):e69546.

4. Levine JA, Schleusner SJ, Jensen MD. Energy expenditure of nonexercise activity. *Am J Clin Nutr.* 2000 Dec;72(6):1451–1454. doi:10.1093/ajcn/72.6.1451.

5. Deguchi Y, Miyazaki K. Anti-hyperglycemic and anti-hyperlipidemic effects of guava leaf extract. *Nutr Metab (Lond).* 2010;7:9. doi:10.1186/1743-7075-7-9.

6. Pigeon WR. Treatment of adult insomnia with cognitive-behavioral therapy. *J Clin Psychol.* 2010 Nov;66(11):1148–1160.

7. Morin CM, Bootzin RR, Buysse DJ, Edinger JD, Espie CA, Lichstein KL. Psychological and behavioral treatment of insomnia: Update of the recent evidence (1998–2004). *Sleep.* 2006 Nov;29(11):1398–1414.

8. Kräuchi K. The thermophysiological cascade leading to sleep initiation in relation to phase of entrainment. *Sleep Med Rev.* 2007 Dec;11(6):439–451.

9. Haghayegh S, Khoshnevis S, Smolensky MH, Diller KR, Castriotta RJ. Before-bedtime passive body heating by warm shower or bath to improve sleep: A systematic review and meta-analysis. *Sleep Med Rev.* 2019 Aug;46:124–135.

10. Ma X, Yue ZQ, Gong ZQ, et al. The effect of diaphragmatic breathing on attention, negative affect and stress in healthy adults. *Front Psychol.* 2017;8:874.

Chapter 14: The Slow Dolphin Power Protocol

1. Ash G, et al. Day-to-day relationships between physical activity and sleep characteristics among people with heart failure and insomnia. *Behav Sleep Med.* 2020.

2. Wehrens SMT, Christou S, Isherwood C, et al. Meal timing regulates the human circadian system. *Curr Biol.* 2017;27(12):1768–1775.e3.

3. Rihm JS, et al. Sleep deprivation selectively unregulates an amygdala-hypothalamic circuit involved in food reward. *J Neurosci.* 2019 Jan 30;39(5):888–899.

4. Deguchi Y, Miyazaki K. Anti-hyperglycemic and anti-hyperlipidemic effects of guava leaf extract. *Nutr Metab (Lond).* 2010;7(9). doi:10.1186/1743-7075-7-9.

5. Pigeon WR. Treatment of adult insomnia with cognitive-behavioral therapy. *J Clin Psychol.* 2010 Nov;66(11):1148–1160.

6. Morin CM, Bootzin RR, Buysse DJ, Edinger JD, Espie CA, Lichstein KL. Psychological and behavioral treatment of insomnia: Update of the recent evidence (1998–2004). *Sleep.* 2006 Nov;29(11):1398–1414.

7. Kräuchi K. The thermophysiological cascade leading to sleep initiation in relation to phase of entrainment. *Sleep Med Rev.* 2007 Dec;11(6):439–451.

8. Haghayegh S, Khoshnevis S, Smolensky MH, Diller KR, Castriotta RJ. Before-bedtime

passive body heating by warm shower or bath to improve sleep: A systematic review and meta-analysis. *Sleep Med Rev.* 2019 Aug;46:124–135.

9. Blanken TF, Benjamins JS, Borsboom D, et al. Insomnia disorder subtypes derived from life history and traits of affect and personality. *Lancet Psychiatry.* 2019 Feb;6(2):151–163.

Chapter 16: Year-Round Energy

1. Lambert GW, Reid C, Kaye DM, Jennings GL, Esler MD. Effect of sunlight and season on serotonin turnover in the brain. *Lancet.* 2002 Dec 7;360(9348):1840–1842.

2. Peiser B. Seasonal affective disorder and exercise treatment: A review. *Biol Rhythm Res.* 2009;40(1):85–97.

3. Gibson EM, Wang C, Tjho S, Khattar N, Kriegsfeld LJ. Experimental "jet lag" inhibits adult neurogenesis and produces long-term cognitive deficits in female hamsters. *PLOS One.* 2010 Dec 1;5(12):e15267.

4. Gunn HE, Troxel WM, Hall MH, Buysse DJ. Interpersonal distress is associated with sleep and arousal in insomnia and good sleepers. *J Psychosom Res.* 2014;76(3):242–248.

5. Szuhany KL, Young A, Mauro C, et al. Impact of sleep on complicated grief severity and outcomes. *Depress Anxiety.* 2020;37:73–80.

6. Tsuno N, Besset A, Ritchie K. Sleep and depression. *J Clin Psychiatry.* 2005;66(10): 1254–1269.

7. Taylor DJ, Lichstein KL, Durrence HH, Reidel BW, Bush AJ. Epidemiology of insomnia, depression, and anxiety. *Sleep.* 2005 Nov;28(11):1457–1464.

8. Toffol E, Merikanto I, Lahti T, Luoto R, Heikinheimo O, Partonen T. Evidence for a relationship between chronotype and reproductive function in women. *Chronobiol Int.* 2013 Jul;30(6):756–765. doi:10.3109/07420528.2012.763043.

9. Ohara T, Nakamura TJ, Nakamura W, et al. Modeling circadian regulation of ovulation timing: Age-related disruption of estrous cyclicity. *Sci Rep.* 2020;10:16767.

10. Kim A, Purse B, Hirsch K, et al. SAT-207 The effects of the menstrual cycle and caloric restriction on sleep in young women. *J Endocr Soc.* 2019;3(Suppl 1):SAT-207.

11. Lee KA, Shaver JF, Giblin EC, Woods NF. Sleep patterns related to menstrual cycle phase and premenstrual affective symptoms. *Sleep.* 1990 Oct;13(5):403–409.

12. Shechter A, Boivin DB. Sleep, hormones, and circadian rhythms throughout the menstrual cycle in healthy women and women with premenstrual dysphoric disorder. *Int J Endocrinol.* 2010;2010:259345.

13. Belanger L, Burt D, Callaghan J, Clifton S, Gleberzon BJ. Anterior cruciate ligament laxity related to the menstrual cycle: An updated systematic review of the literature. *J Can Chiropr Assoc.* 2013;57(1):76–86.

Chapter 17: Energy and Your Health

1. Popkin BM, D'Anci KE, Rosenberg IH. Water, hydration, and health. *Nutr Rev.* 2010;68(8):439–458.

2. Dimitrov S, Lange T, Gouttefangeas C, et al. Gα$_s$-coupled receptor signaling and sleep regulate integrin activation of human antigen-specific T cells. *J Exp Med.* 2019 Mar 4;216(3):517–526.

3. Prather AA, Janicki-Deverts D, Hall MH, Cohen S. Behaviorally assessed sleep and susceptibility to the common cold. *Sleep*. 2015 Sep 1;38(9):1353–1359. doi:10.5665/sleep.4968.

4. Molina E, Sanabria D. Effects of chronotype and time of day on the vigilance decrement during simulated driving. *Accid Anal Prev*. 2014;67C:113–118. doi:10.1016/j.aap.2014.02.020.

5. Gao B, Dwivedi S, Milewski MD, Cruz AI Jr. Chronic lack of sleep is associated with increased sports injury in adolescents: A systematic review and meta-analysis. *Orthop J Sports Med*. 2019;7(3 Suppl):2325967119S00132. doi:10.1177/2325967119S00132.

6. Lima LV, Abner TSS, Sluka KA. Does exercise increase or decrease pain? Central mechanisms underlying these two phenomena. *J Physiol*. 2017;595(13):4141–4150.

7. Geneen LJ, Moore RA, Clarke C, Martin D, Colvin LA, Smith BH. Physical activity and exercise for chronic pain in adults: An overview of Cochrane Reviews. *Cochrane Database Syst Rev*. 2017 Jan 14;1(1):CD011279.

Index

Note: Italic page numbers refer to illustrations and charts.

About the Authors

Michael Breus, PhD, is a clinical psychologist and both a diplomate of the American Board of Sleep Medicine and a fellow of the American Academy of Sleep Medicine. Dr. Breus has been in practice for twenty-three years, has been featured on *The Dr. Oz Show* more than forty times, writes regularly for *Psychology Today,* and was named the top sleep specialist in California by *Reader's Digest.* He is the author of three Amazon bestsellers: *The Power of When* (2016), *The Sleep Doctor's Diet Plan* (2011), and *Beauty Sleep* (2006).

thesleepdoctor.com

Twitter @thesleepdoctor | Facebook.com/thesleepdoctor

Instagram @thesleepdoctor

Stacey Griffith is the founding senior master instructor at SoulCycle, the bestselling author of *Two Turns from Zero,* and a co-host of the popular podcast *The Way* with Sarah Wragge. Griffith has been featured in national media outlets, including the *New York Times,* the *Wall Street Journal, Vogue, People, Town and Country, Vanity Fair, Self, Shape, Women's Health, New York,* and the *New York Post,* and she is a highly sought-after motivational speaker.

staceygnyc.com

Twitter @staceygnyc | Instagram @staceygnyc